Lecture Notes in Computer Science

Vol. 324: M.P. Chytil, L. Janiga, V. Koubek (Eds.), Mathematical Foundations of Computer Science 1988. Proceedings. IX, 562 pages. 1988.

Vol. 325: G. Brassard, Modern Cryptology. VI, 107 pages. 1988.

Vol. 326: M. Gyssens, J. Paredaens, D. Van Gucht (Eds.), ICDT '88. 2nd International Conference on Database Theory. Proceedings, 1988. VI, 409 pages. 1988.

Vol. 327: G.A. Ford (Ed.), Software Engineering Education. Proceedings, 1988. V, 207 pages. 1988.

Vol. 328: R. Bloomfield, L. Marshall, R. Jones (Eds.), VDM '88. VDM – The Way Ahead. Proceedings, 1988. IX, 499 pages. 1988.

Vol. 329: E. Börger, H. Kleine Büning, M.M. Richter (Eds.), CSL '87. 1st Workshop on Computer Science Logic. Proceedings, 1987. VI, 346 pages. 1988.

Vol. 330: C.G. Günther (Ed.), Advances in Cryptology – EUROCRYPT '88. Proceedings, 1988. XI, 473 pages. 1988.

Vol. 331: M. Joseph (Ed.), Formal Techniques in Real-Time and Fault-Tolerant Systems. Proceedings, 1988. VI, 229 pages. 1988.

Vol. 332: D. Sannella, A. Tarlecki (Eds.), Recent Trends in Data Type Specification. V, 259 pages. 1988.

Vol. 333: H. Noltemeier (Ed.), Computational Geometry and its Applications. Proceedings, 1988. VI, 252 pages. 1988.

Vol. 334: K.R. Dittrich (Ed.), Advances in Object-Oriented Database Systems. Proceedings, 1988. VII, 373 pages. 1988.

Vol. 335: F.A. Vogt (Ed.), CONCURRENCY 88. Proceedings, 1988. VI, 401 pages. 1988.

Vol. 336: B.R. Donald, Error Detection and Recovery in Robotics. XXIV, 314 pages. 1989.

Vol. 337: O. Günther, Efficient Structures for Geometric Data Management. XI, 135 pages. 1988.

Vol. 338: K.V. Nori, S. Kumar (Eds.), Foundations of Software Technology and Theoretical Computer Science. Proceedings, 1988. IX, 520 pages. 1988.

Vol. 339: M. Rafanelli, J.C. Klensin, P. Svensson (Eds.), Statistical and Scientific Database Management, IV SSDBM. Proceedings, 1988. IX, 454 pages. 1989.

Vol. 340: G. Rozenberg (Ed.), Advances in Petri Nets 1988. VI, 439 pages. 1988.

Vol. 341: S. Bittanti (Ed.), Software Reliability Modelling and Identification. VII, 209 pages. 1988.

Vol. 342: G. Wolf, T. Legendi, U. Schendel (Eds.), Parcella '88. Proceedings, 1988. 380 pages. 1989.

Vol. 343: J. Grabowski, P. Lescanne, W. Wechler (Eds.), Algebraic and Logic Programming. Proceedings, 1988. 278 pages. 1988.

Vol. 344: J. van Leeuwen, Graph-Theoretic Concepts in Computer Science. Proceedings, 1988. VII, 459 pages. 1989.

Vol. 345: R.T. Nossum (Ed.), Advanced Topics in Artificial Intelligence. VII, 233 pages. 1988 (Subseries LNAI).

Vol. 346: M. Reinfrank, J. de Kleer, M.L. Ginsberg, E. Sandewall (Eds.), Non-Monotonic Reasoning. Proceedings, 1988. XIV, 237 pages. 1989 (Subseries LNAI).

Vol. 347: K. Morik (Ed.), Knowledge Representation and Organization in Machine Learning. XV, 319 pages. 1989 (Subseries LNAI).

Vol. 348: P. Deransart, B. Lorho, J. Maluszyński (Eds.), Programming Languages Implementation and Logic Programming. Proceedings, 1988. VI, 299 pages. 1989.

Vol. 349: B. Monien, R. Cori (Eds.), STACS 89. Proceedings, 1989. VIII, 544 pages. 1989.

Vol. 350: A. Törn, A. Žilinskas, Global Optimization. X, 255 pages. 1989.

Vol. 351: J. Díaz, F. Orejas (Eds.), TAPSOFT '89. Volume 1. Proceedings, 1989. X, 383 pages. 1989.

Vol. 35[illegible] 2. Proceeding[illegible]

Vol. 353: [illegible] ming. X, 250 page[illegible]

Vol. 354: [illegible] [illegible]berg (Eds.), Linear Tim[illegible] Time and Partial Order in Logics and Models for Concurrency. VIII, 713 pages. 1989.

Vol. 355: N. Dershowitz (Ed.), Rewriting Techniques and Applications. Proceedings, 1989. VII, 579 pages. 1989.

Vol. 356: L. Huguet, A. Poli (Eds.), Applied Algebra, Algebraic Algorithms and Error-Correcting Codes. Proceedings, 1987. VI, 417 pages. 1989.

Vol. 357: T. Mora (Ed.), Applied Algebra, Algebraic Algorithms and Error-Correcting Codes. Proceedings, 1988. IX, 481 pages. 1989.

Vol. 358: P. Gianni (Ed.), Symbolic and Algebraic Computation. Proceedings, 1988. XI, 545 pages. 1989.

Vol. 359: D. Gawlick, M. Haynie, A. Reuter (Eds.), High Performance Transaction Systems. Proceedings, 1987. XII, 329 pages. 1989.

Vol. 360: H. Maurer (Ed.), Computer Assisted Learning – ICCAL '89. Proceedings, 1989. VII, 642 pages. 1989.

Vol. 361: S. Abiteboul, P.C. Fischer, H.-J. Schek (Eds.), Nested Relations and Complex Objects in Databases. VI, 323 pages. 1989.

Vol. 362: B. Lisper, Synthesizing Synchronous Systems by Static Scheduling in Space-Time. VI, 263 pages. 1989.

Vol. 363: A.R. Meyer, M.A. Taitslin (Eds.), Logic at Botik '89. Proceedings, 1989. X, 289 pages. 1989.

Vol. 364: J. Demetrovics, B. Thalheim (Eds.), MFDBS 89. Proceedings, 1989. VI, 428 pages. 1989.

Vol. 365: E. Odijk, M. Rem, J.-C. Syre (Eds.), PARLE '89. Parallel Architectures and Languages Europe. Volume I. Proceedings, 1989. XIII, 478 pages. 1989.

Vol. 366: E. Odijk, M. Rem, J.-C. Syre (Eds.), PARLE '89. Parallel Architectures and Languages Europe. Volume II. Proceedings, 1989. XIII, 442 pages. 1989.

Vol. 367: W. Litwin, H.-J. Schek (Eds.), Foundations of Data Organization and Algorithms. Proceedings, 1989. VIII, 531 pages. 1989.

Vol. 368: H. Boral, P. Faudemay (Eds.), IWDM '89, Database Machines. Proceedings, 1989. VI, 387 pages. 1989.

Vol. 369: D. Taubner, Finite Representations of CCS and TCSP Programs by Automata and Petri Nets. X. 168 pages. 1989.

Vol. 370: Ch. Meinel, Modified Branching Programs and Their Computational Power. VI, 132 pages. 1989.

Vol. 371: D. Hammer (Ed.), Compiler Compilers and High Speed Compilation. Proceedings, 1988. VI, 242 pages. 1989.

Vol. 372: G. Ausiello, M. Dezani-Ciancaglini, S. Ronchi Della Rocca (Eds.), Automata, Languages and Programming. Proceedings, 1989. XI, 788 pages. 1989.

Vol. 373: T. Theoharis, Algorithms for Parallel Polygon Rendering. VIII, 147 pages. 1989.

Vol. 374: K.A. Robbins, S. Robbins, The Cray X-MP/Model 24. VI, 165 pages. 1989.

Vol. 375: J.L.A. van de Snepscheut (Ed.), Mathematics of Program Construction. Proceedings, 1989. VI, 421 pages. 1989.

Vol. 376: N.E. Gibbs (Ed.), Software Engineering Education. Proceedings, 1989. VII, 312 pages. 1989.

Vol. 377: M. Gross, D. Perrin (Eds.), Electronic Dictionaries and Automata in Computational Linguistics. Proceedings, 1987. V, 110 pages. 1989.

Vol. 378: J.H. Davenport (Ed.), EUROCAL '87. Proceedings, 1987. VIII, 499 pages. 1989.

Lecture Notes in Computer Science

Edited by G. Goos and J. Hartmanis

426

Niels Houbak

SIL – a Simulation Language

User's Guide

Springer-Verlag

Berlin Heidelberg New York London Paris Tokyo Hong Kong

Author

Niels Houbak
Laboratory for Energetics, Build. 403
Technical University of Denmark
DK-2800 Lyngby, Denmark

CR Subject Classification (1987): I.6.2

ISBN 3-540-52497-5 Springer-Verlag Berlin Heidelberg New York
ISBN 0-387-52497-5 Springer-Verlag New York Berlin Heidelberg

Printed in Germany

Printing and binding: Druckhaus Beltz, Hemsbach/Bergstr.
2145/3140-543210 – Printed on acid-free paper

Preface

The development of SIL started in the early spring of 1981 with a formal description of its syntax, some of its semantics and the facilities it should include. These considerations were the basis for an application to the Danish Technical Research Council. During the two year project most of the compiler and all the vital parts of the run-time system were implemented as two separate programs. After those two years the SIL system was capable of solving most of the problems it was originally designed for, in batch mode, with output either as tables or line printer plots. All development until then had been on IBM machines running VM/CMS using the PASCAL/VS compiler. SIL was also installed on an MVS machine using an ALGOL W version of the run-time system.

Until 1985 SIL developed slowly but with a growing interest from users; several bugs were discovered due to user reports and removed. At that time, the SIL system was merged into one program (model interpretation) and transferred to an IBM PC. To fully exploit the PC a model editor was incorporated together with the graphics and the interactive command system. In 1988 the status of the system is that it runs on most IBM (or compatible) PC's and it can be used easily by inexperienced users to simulate a wide range of problems.

There are several sources of inspiration for the μSIL system as it is today. First of all, classic MATLAB (written in FORTRAN by C. Moler (1980)) convinced me that it was at all possible to make such a program within a reasonable time; and second, SIMNON (Elmquist (1975)) gave some guidelines for what to do and what not to do.

The μSIL system is a closed system; that is, it does not allow any part of a simulation model to be given externally, for example as a FORTRAN subroutine. The system consists of a compiler, a line editor, a model interpreter, a graphic command system, and numerical software for solving the model. The SIL language is ALGOL or PASCAL-like. Its compiler is a two-pass top down compiler (inspired by Wirth (1976)) which generates intermediate code suitable for the interpreter. The interpreter takes the model in Reverse Polish notation and evaluates the expressions using a stack.

This users guide is intended both for the 'first-time' user and for more experienced users. Chapters 1, 2, and 3 are a very brief introduction to how to install the μSIL system and how to run simple models. Chapter 4 is a general description of what simulation is and how simulation models are built. Chapter 5 concerns the numerical methods used for solving the mathematical models. Chapter 6 is a systematic description of the SIL language used for building simulation models in the SIL system. Chapter 7 gives a complete description of the SIL system (except the language), and Chapter 8 concludes with examples of the use of the system. Chapters 6 and 7 can be used as a reference manual for the system. The four appendices contain

A) a full syntax of the language,
B) a listing of the on-line help file,
C) all error codes issued by the system, and
D) the documentation of the integration routine used.

Though I have done most of the programming of the system myself, many of my colleagues during my seven years with SIL have greatly influenced different parts of it. First of all I would like to thank my former supervisor Docent Per G. Thomsen with whom many details have been discussed; SIL would never have been the same without him. Also, I owe many thanks to Peter Høj who for a long period shared a strong interest for the SIL system. I would also like to take this opportunity to express my gratitude to all the staff at the Institute for Numerical Analysis, Technical University of Denmark for the very fine working conditions they gave me until I moved in 1984 to the Laboratory for Energetics.

I would also like to thank some of the early and/or active users at the university (DTH/DIA). They are O. Jensen, DIA-E; P. Danig, LKT-DTH; O.B. Nielsen, LFE-DTH; G. Winkler, Fachhochschule Flensburg; E. Hendricks, IMSOR-DTH; S.C. Sorenson, LFE-DTH; L. Kristensen, LFE-DTH and several master students at LFE using SIL in their final thesis project. Solving their problems (and sometime finding bugs in the SIL system) has been a big help for me and a great source of inspiration for new facilities in the system.

Finally, I would like to thank the rest of the staff at the Laboratory for Energetics (DTH) for their support and encouragement, including especially J. Reffstrup and Professor B. Qvale. Last but not least I would like to thank my wife Pia, because without her patience and understanding of my working overtime and at home, there would not have been any SIL system as it is today.

This manual has a Danish predecessor from 1982 mainly describing the SIL language. A part of that manual was translated into English by Peter Høj and this in revised form has become Chapter 6 of this manual. In the preparation of this manuscript Elbert Hendriks and Spencer C. Sorenson have been very helpful both concerning the content and my spelling; however, they are not to blame for any remaining spelling errors.

Throughout this manual several trademarks registered by major companies have been used with all respect. A floppy disc containing the μSIL system and several of the examples from this manual can be obtained from the author.

Copenhagen, January 1990 Niels Houbak

Table of contents

1. General Introduction.

This manual is a complete description of the μSIL simulation system. The system is a closed whole-in-one system; it has its own language, the SIL language, used for building models of the systems to be simulated; it has its own graphic command system for running the models and presenting the results graphically; and it has a built-in interactive editor with on-line model syntax analysis.

To some people the word simulation is a loaded word. In this context we use 'simulation' as a synonym for the process of making a (mathematical) model of a 'real-world' system and solving (one way or the other) this model. Inconsistently, we will sometimes use the phrase 'to simulate' about the solution process. It is beyond the scope of this manual to go into details about how to **build** mathematical models of physical systems for two reasons,

1) several textbooks already exist on the market, and
2) in each area of application there normally is a strong tradition for how this is done and these traditions are different from one field to another.

μSIL can be used for solving first or higher order systems of non-linear ordinary differential or difference equations coupled with systems of non-linear algebraic equations. It can handle discontinuities (switching between several states) and combined discrete/continuous systems. μSIL can not be used for solving partial differential equations. The integrated output processing can plot results versus time, make phase plane plots, and/or make user-defined output tables.

The SIL language is based on an 'ALGOL/PASCAL'-like description. All variables must be declared before being used. In contrast to most programming languages, the type of a variable corresponds to its application and not to its internal representation; all results are calculated in 'double precision' (approximately 8 bytes reals). The types of (simulation-) variables are VARIABLE, DERIVATIVE, and PARAMETER. Other type definitions are SWITCH, SAMPLETIME, DISCRETE,

TIME, and MACRO. The language has standard assignment statements, and implicit assignment statements are used for residual equations. Nestable IF THEN ELSE statements can be used to define discontinuities. Several predefined internal global variables are used to control the integration process. A powerful internal automatic syntax and semantic control of the model reduces the possibility of making errors.

A large effort has been invested in μSIL to ensure high efficiency, reliability and robustness. The systems of ODEs are solved by STRIDE, a variable order, variable stepsize, singly diagonally implict Runge-Kutta method due to Burrage, Butcher, and Chipman (1980). The system of implicit algebraic equations is solved by a Quasi-Newton method. Discontinuities are located in time and the integration is restarted after each passage. Each variable can have a validity range; if any range is violated the integration of the problem stops.

The μSIL system is entirely written in PASCAL and on PC's it is compiled using TURBO PASCAL. The use of PASCAL has made it easy to write the compiler that analyses the model, and with a compile rate of 400 lines of SIL program per minute on an IBM PC/XT-286 it is reasonably efficient. The Van der Pol equation (a 2. order non-linear ordinary differential equation) is solved in less than 30 seconds dependent on the quantity of output requested. This time is of the same order as the time used by most FORTRAN integration subroutines for the same problem.

There is one very important thing that should be mentioned here; no matter the quality of the simulation software you use, if the model is wrong, the results will be wrong. Garbage in - garbage out. The μSIL system can produce one (of several) mathematical solutions to the model; the user is responsible for verifying the results.

1.1. Hardware Requirements.

In order to use all the facilities of the μSIL system your PC must have a certain minimum hardware configuration. In this chapter we will describe the hardware facilities required by the μSIL system.

Your PC must be of XT/AT type (IBM or compatible) using either MS-DOS or PC-DOS, version 2.0 or later. If your PC only has 512 Kbytes of RAM it is sufficient but with the larger storage available larger models can be solved. The μSIL system itself takes approximately 220 Kbytes plus 100 Kbytes for data. All storage not used for other purposes, is used by the μSIL system for an internal copy of the model and all the output points. Since the DOS system normally occupies some 60 Kbytes we recommend that your PC should have all the 640 Kbytes of RAM that DOS allows. The μSIL system is not currently capable of using storage beyound the 640 Kbytes boundary.

Your PC system must have a math co-processor (80x87). The μSIL system cannot be executed unless it's there. The reason for this is that during a simulation the μSIL system does many floating point calculations; without a math co-processor the computing time increases with a factor of 3 to 4 and the accuracy of the results will be almost halved. Therefore, the μSIL system is compiled in such a way that it requires the presence of a math co-processor.

If you want to see the results of your simulations in a convenient way your PC must have a monitor graphics display board. The μSIL system supports several graphics systems, so if you have graphics on your PC compatible with any of those you can fully exploit the features of the μSIL system. The graphics cards currently supported are the standard IBM graphics, CGA, EGA, VGA, and MCGA; there is also support for the Hercules monocrome graphics card and for the Olivetti (AT&T 400) built-in graphics card. Generally, the graphics mode for all the cards are chosen by μSIL to give high resolution and few (two) colours. The graphics card is not a requirement since the μSIL system can produce results either as columns of numbers or as rough 'line-printer' plots. A nice 'companion' to a graphics card is software that enables you to dump the (graphic) screen to an attatched printer. There is no 'hardcopy' facillity built into the μSIL system nor is there any possibility of directing the output to a plotting device.

The μSIL system is normally distributed on floppy disk(s). It is possible to run the system directly from those disk(s), but in order to load the μSIL system faster it is recommended that you have a hard disk on your PC. The system does NOT rely on a fast hard disk for either overlay or temporary storage of data.

2. Installation of the μSIL System.

Now we will show how you can install the μSIL system on your harddisk (for convenience we assume that you have a harddisk). It is not the only way of installing the system, but it is a good way. For those who might be interested in a more detailed description of the system structure and the use of files, perhaps for using the μSIL system in a network environment, we refer to chapter 7. The following description assumes that you are familiar with the DOS directory concept.

Directly under the root directory of you harddisk create a subdirectory named 'SIL'. Copy all the files from the distribution diskette into this subdirectory. There is no special INSTALL procedure. We will shortly mention some of the most important files that you must copy. It is essential that all the files 'SIL.*' and '*.BGI' are copied all together to the same and only subdirectory.

SIL.EXE This file contains the executeable part of the μSIL system.

SIL.ERM Contains all the error messages that can be written by the μSIL system. You can print this file, but be careful not to edit the file with a 'tab setting' editor.

SIL.HLP All HELP messages for the interactive editor and the command processor are stored in this file. Print the file if you want to. Appendix B contains a listing of this file.

SIL.ACC Accounting information. The μSIL system requires WRITE access to this file. It is updated each time μSIL is called. This one-line file also contains vital information for the system to function. <u>WARNING</u>: IF THIS FILE IS SUBJECT TO ANY UN-AUTHORIZED CHANGES THE μSIL SYSTEM WILL NEVER BE ABLE TO USE IT AGAIN AND THE μSIL SYSTEM WILL NOT RUN.

SIL.DEV The first line of this file must contain the PATH to the directory where you have stored the μSIL system. A copy of this file should be present in all directories where you have SIL models. If this file is not present, μSIL will generate it by searching the current drive for a file named SIL.EXE. You may change this file if you need. Below we give an example of the content of such a file:

```
C:\SIL\
```

SIL.LGO If this file is on the distribution disk it contains licensee information. The file will be typed on the screen each time μSIL is activated. The file can be changed without serious consequences.

SILREAD.ME This file contains supplementary information about the system and some last minute updates (if any). This file should be printed out when the system is started up the first time.

SILERROR.DOC Known bugs, deficiencies and comments can be found in this file. Also, when bugs have been corrected this file will be used for notification. This file is also used for informing you about what can be expected of new features in the system. In case you have additional wishes please don't hesitate to inform the author.

*.BGI BORLAND graphic device drivers used by the μSIL system. If these files are not present on the distribution diskette they are included in SIL.EXE .

*.SIL Sample problems that should run correctly on your μSIL system. These problems exercise most of the SIL language and since they are reasonably well documented with comments, you might be able to learn the SIL language by reading them.

If you want all your SIL models to be in the SIL subdirectory you are ready to run the system. If you want your SIL models in separate subdirectories you must copy the file SIL.DEV into all of those subdirectories. It is essential that files used for storing SIL models have the extension '.SIL'. As explained in chapter 7, you may have one subdirectory as the current directory and run a model in another subdirectory.

You can use almost any editor for making SIL models. It is essential though that the editor can store the model on disk in DOS text mode; that is, the only special character recognized by μSIL is HT (the Horizontal Tab character), and the length of any line read is limited to 80 characters. If you can TYPE (the DOS command) the file and get it nicely on your screen, it should be readable by μSIL. If the file produced by your favorite editor is unreadable by μSIL you can always use the built-in line-oriented editor. This facility will be described later but it is important to notice though, that this editor can only store correct SIL models on disk.

3. Getting Started.

Now you should be ready to run the μSIL system the very first time. Having your newly created subdirectory 'SIL' as the current directory you may issue the DOS command DIR *.SIL and get a list of all the sample models that are currently available. One of these files (supplied with μSIL) has the name 'VANPOL.SIL', and contains a SIL model of a second order non-linear differential equation named the Van der Pol equation. You can print the file to see what it looks like or you can use your favorite editor to look at it. Figure 3.1 shows the contents of this file.

```
$TITLE=   The first simple problem

(*   model, describing the simple
     Van der Pol equation              *)
BEGIN

(*   declarations      *)

VARIABLE Y(2);          (* variable  Y  with initial value  *)
PARAMETER EPS(10);
DERIVATIVE DY(Y)(0),    (* variable  DY  with initial value *)
           D2Y(DY);     (* second derivative of Y           *)
TIME T(0:20);           (* model time (indep. variable      *)

(*   changing default values of predeclared
     variables for run control                              *)

RELERROR:=1'-4;   ABSERROR:=1.'-4;

(*   the differential equation         *)

D2Y := EPS*(1-Y*Y)*DY - Y;

(*   output specification      *)

PRINT(80,Y,DY);        (* time plot of  Y  and  DY      *)
PRINT(1024,Y(DY));     (* phase plane plot Y versus DY *)
WRITE(Y,DY)            (* tabulate the solution         *)

END.
```

Figure 3.1: A SIL model of the Van der Pol equation.

We will later return to a description of the SIL language. This description will (hopefully) answer most of the more detailed questions you have when seeing a model like this one. Notice though, the text typed between (* and *) is treated as a comment, and although it is very useful to have comments in a model it is not necessary.

When you want to run the model with the μSIL system, simply type at the DOS prompt the following command: SIL VANPOL or if you prefer: SIL VANPOL.SIL

First, DOS loads the μSIL system into memory and relinquishes the control to it. μSIL scans the command line parameter VANPOL and starts reading and analysing the SIL model stored in the file VANPOL.SIL. After a few seconds, μSIL is ready to solve the problem and when doing so it updates a screen monitor line with the number of steps completed and the actual model time (not CPU-time).

After some time (a minute or so) the simulation will finish and the DOS prompt will appear on the screen. If the simulation is done, where are then the results? Make a list of the files in the directory using the DIR command. A new file named 'VANPOL.LST' appears in the list. This file has been generated by μSIL and it contains all the results; print this file, look at it, and see how the output from μSIL is organized. There is a listing of the SIL model as read from the file 'VANPOL.SIL' with line numbers added. Then there are some statistics from the simulation phase, and finally you will find the results of the simulation presented as specified in the output statements of the SIL model. The following pages show what this file should look like.

```
SIL VERSION 2.0 (880419)
N. HOUBAK   86-10-27            The first simple problem                    88-08-03  08:06:51  PAGE 01

(*
     model, describing the simple
     Van der Pol equation           *)
BEGIN

(*   declarations    *)

VARIABLE Y(2);    (* variable  Y  with initial value *)
PARAMETER EPS(10);
DERIVATIVE DY(Y)(0),(* variable  DY  with init. val. *)
           D2Y(DY); (* second derivative of Y        *)
TIME T(0:20);        (* model time (indep. variable   *)

(*   changing default values of predeclared
     variables for run control                       *)

RELERROR:=1'-4;   ABSERROR:=1.'-4;

(*   the differential equation         *)

D2Y := EPS*(1-Y*Y)*DY - Y;

(*   output specification     *)
PRINT(80,Y,DY);      (* time plot of  Y  and  DY     *)
PRINT(1024,Y(DY));  (* phase plane plot Y versus DY *)
WRITE(Y,DY)          (* tabulate the solution        *)

END.

  2.42  SECONDS IN COMPILATION

MODEL CONSISTS OF :
   1 PARAMETERS
   2 DYNAMIC VARIABLES / DERIVATIVES



SIMULATION STATISTICS:

   NUMBER OF ACCEPTED STEPS       :  408
   NUMBER OF REJECTED STEPS       :   15
   NUMBER OF DIVERGENCES          :   86
   NUMBER OF FUNCTION CALLS       : 2969
   TOTAL NUMBER OF FUNCTION CALLS : 2971


SIMULATION OPTIONS USED:

   DEBUG              :    1
   METHOD             :  119
   MAXORDER           :   10
   MAXCPU             :  0.0
   INITIAL TIME       : 0.000E+0000
   FINAL TIME         : 2.000E+0001
   MAXIMUM STEPSIZE   : 0.000E+J000
   INITIAL STEPSIZE   : 0.000E+0000
   ABSERROR           : 1.000E-0004
   RELERROR           : 1.000E-0004


VALUES OBTAINED DURING SIMULATION:

   MAXORDER           :    4
   MAXIMUM STEPSIZE   : 2.407E-0001
   MINIMUM STEPSIZE   : 1.953E-0004


PARAMETER VALUES :
EPS          =  1.00000E+0001


INITIAL VALUES :
Y            =  2.00000E+0000 DY           =  0.00000E+0000


FINAL VALUES FOR DYNAMIC VARIABLES:
Y            =  1.93367E+0000 DY           = -7.04272E-0002
```

S I L - P L O T   O F   R E S U L T S

PRINTING Y AND DY VERSUS T

```
Y    = -2.50E+00 -2.00E+00 -1.50E+00 -1.00E+00 -5.00E-01  0.00E+00  5.00E-01  1.00E+00  1.50E+00  2.00E+00  2.50E+00
DY   = -1.50E+01 -1.20E+01 -9.00E+00 -6.00E+00 -3.00E+00  0.00E+00  3.00E+00  6.00E+00  9.00E+00  1.20E+01  1.50E+01
 0.00E+00 +----+----+----+----+----+----+----+----+----+----D----+----+----+----+----+----+----+----+----Y----+----+
 2.50E-01 !         !         !         !         !         D         !         !         !         Y         !
 5.00E-01 !         !         !         !         !         D         !         !         !        YY         !
 7.50E-01 !         !         !         !         !         D         !         !         !        Y!         !
 1.00E+00 !         !         !         !         !         D         !         !         !        Y!         !
 1.25E+00 !         !         !         !         !         D         !         !         !       Y !         !
 1.50E+00 !         !         !         !         !         D         !         !         !       Y !         !
 1.75E+00 !         !         !         !         !         D         !         !         !      YY !         !
 2.00E+00 !         !         !         !         !         D         !         !         !      Y  !         !
 2.25E+00 !         !         !         !         !         D         !         !         !      Y  !         !
 2.50E+00 +----+----+----+----+----+----+----+----+----+----D----+----+----+----+----+----+----+YY--+----+----+
 2.75E+00 !         !         !         !         !         D         !         !         !     Y   !         !
 3.00E+00 !         !         !         !         !         D         !         !         !    YY   !         !
 3.25E+00 !         !         !         !         !         D         !         !         !    Y    !         !
 3.50E+00 !         !         !         !         !         D         !         !         !    Y    !         !
 3.75E+00 !         !         !         !         !         D         !         !         !   YY    !         !
 4.00E+00 !         !         !         !         !         D         !         !         !   Y     !         !
 4.25E+00 !         !         !         !         !         D         !         !         !  YY     !         !
 4.50E+00 !         !         !         !         !         D         !         !         !  Y      !         !
 4.75E+00 !         !         !         !         !         D         !         !         ! YY      !         !
 5.00E+00 +----+----+----+----+----+----+----+----+----+----D----+----+----+----+----+----+-Y--+----+----+----+
 5.25E+00 !         !         !         !         !        DD         !         !         !YY       !         !
 5.50E+00 !         !         !         !         !        DD         !         !         !Y        !         !
 5.75E+00 !         !         !         !         !        DD         !         !         YY        !         !
 6.00E+00 !         !         !         !         !        DD         !         !         Y         !         !
 6.25E+00 !         !         !         !         !        DD         !         !        Y!         !         !
 6.50E+00 !         !         !         !         !        DD         !         !       YY!         !         !
 6.75E+00 !         !         !         !         !        DD         !         !       Y !         !         !
 7.00E+00 !         !         !         !         !        DD         !         !      Y  !         !         !
 7.25E+00 !         !         !         !         !        DD         !         !     Y   !         !         !
 7.50E+00 +----+----+----+----+----+----+----+----+----+---DD----+----+----+----+----Y----+----+----+----+----+
 7.75E+00 !         !         !         !         !        D!         !         !   Y     !         !         !
 8.00E+00 !         !         !         !         !        D!         !         !  Y      !         !         !
 8.25E+00 !         !         !         !         !        D!         !         !YY       !         !         !
 8.50E+00 !         !         !         !         !       D !         !        Y!         !         !         !
 8.75E+00 !         !         !         !         !    D    !         !    Y    !         !         !         !
 9.00E+00 !         !     DD  !         !         !    Y    !         !         !         !         !         !
 9.25E+00 !         Y         !         !         !        DD         !         !         !         !         !
 9.50E+00 !         Y         !         !         !         D         !         !         !         !         !
 9.75E+00 !         Y         !         !         !         D         !         !         !         !         !
 1.00E+01 +----+----YY---+----+----+----+----+----+----+----D----+----+----+----+----+----+----+----+----+----+
 1.03E+01 !         !Y        !         !         !         D         !         !         !         !         !
 1.05E+01 !         !Y        !         !         !         D         !         !         !         !         !
 1.08E+01 !         ! Y       !         !         !         D         !         !         !         !         !
 1.10E+01 !         ! Y       !         !         !         D         !         !         !         !         !
 1.13E+01 !         ! YY      !         !         !         D         !         !         !         !         !
 1.15E+01 !         !  Y      !         !         !         D         !         !         !         !         !
 1.18E+01 !         !  Y      !         !         !         D         !         !         !         !         !
 1.20E+01 !         !  YY     !         !         !         D         !         !         !         !         !
 1.23E+01 !         !   Y     !         !         !         D         !         !         !         !         !
 1.25E+01 +----+----+---YY----+----+----+----+----+----+----D----+----+----+----+----+----+----+----+----+----+
 1.28E+01 !         !    Y    !         !         !         D         !         !         !         !         !
 1.30E+01 !         !    Y    !         !         !         D         !         !         !         !         !
 1.33E+01 !         !    YY   !         !         !         D         !         !         !         !         !
 1.35E+01 !         !     Y   !         !         !         D         !         !         !         !         !
 1.38E+01 !         !     YY  !         !         !         D         !         !         !         !         !
 1.40E+01 !         !      Y  !         !         !         D         !         !         !         !         !
 1.43E+01 !         !      YY !         !         !         D         !         !         !         !         !
 1.45E+01 !         !       Y !         !         !         D         !         !         !         !         !
 1.48E+01 !         !       YY!         !         !         DD        !         !         !         !         !
 1.50E+01 +----+----+----+---Y+----+----+----+----+----+----DD---+----+----+----+----+----+----+----+----+----+
 1.53E+01 !         !        YY         !         !         DD        !         !         !         !         !
 1.55E+01 !         !         Y         !         !         DD        !         !         !         !         !
 1.58E+01 !         !         !Y        !         !         DD        !         !         !         !         !
 1.60E+01 !         !         !YY       !         !         DD        !         !         !         !         !
 1.63E+01 !         !         ! Y       !         !         DD        !         !         !         !         !
 1.65E+01 !         !         !  Y      !         !         DD        !         !         !         !         !
 1.68E+01 !         !         !   Y     !         !         DD        !         !         !         !         !
 1.70E+01 !         !         !    Y    !         !         DD        !         !         !         !         !
 1.73E+01 !         !         !     Y   !         !         !D        !         !         !         !         !
 1.75E+01 +----+----+----+----+----+-Y--+----+----+----+----+D---+----+----+----+----+----+----+----+----+----+
 1.78E+01 !         !         !       YY!         !         !D        !         !         !         !         !
 1.80E+01 !         !         !         !Y        !         ! D       !         !         !         !         !
 1.83E+01 !         !         !         !    YY   !         !    D    !         !         !         !         !
 1.85E+01 !         !         !         !         !         !     Y   !         !         !   DD    !         !
 1.88E+01 !         !         !         !         !         DD        !         !         !         Y         !
 1.90E+01 !         !         !         !         !         D         !         !         !         Y         !
 1.93E+01 !         !         !         !         !         D         !         !         !         Y         !
 1.95E+01 !         !         !         !         !         D         !         !         !        YY         !
 1.98E+01 !         !         !         !         !         D         !         !         !        Y!         !
 2.00E+01 +----+----+----+----+----+----+----+----+----+----D----+----+----+----+----+----+----+---Y+----+----+

   >> T <<
```

S I L - P L O T   O F   R E S U L T S

PRINTING Y VERSUS DY

```
Y    = -2.50E+00 -2.00E+00 -1.50E+00 -1.00E+00 -5.00E-01  0.00E+00  5.00E-01  1.00E+00  1.50E+00  2.00E+00  2.50E+00
-2.00E+01 +----+----+----+----+----+----+----+----+----+----+----+----+----+----+----+----+----+----+----+----+
-1.96E+01 !         !         !         !         !         !         !         !         !         !         !
-1.92E+01 !         !         !         !         !         !         !         !         !         !         !
-1.88E+01 !         !         !         !         !         !         !         !         !         !         !
-1.84E+01 !         !         !         !         !         !         !         !         !         !         !
-1.80E+01 !         !         !         !         !         !         !         !         !         !         !
-1.76E+01 !         !         !         !         !         !         !         !         !         !         !
-1.72E+01 !         !         !         !         !         !         !         !         !         !         !
-1.68E+01 !         !         !         !         !         !         !         !         !         !         !
-1.64E+01 !         !         !         !         !         !         !         !         !         !         !
-1.60E+01 +----+----+----+----+----+----+----+----+----+----+----+----+----+----+----+----+----+----+----+----+
-1.56E+01 !         !         !         !         !         !         !         !         !         !         !
-1.52E+01 !         !         !         !         !         !         !         !         !         !         !
-1.48E+01 !         !         !         !         !         !         !         !         !         !         !
-1.44E+01 !         !         !        Y!         !         !         !         !         !         !         !
-1.40E+01 !         !         !        Y!         !         !         !         !         !         !         !
-1.36E+01 !         !         !         !   YY    !         !         !         !         !         !         !
-1.32E+01 !         !         !  YY     !         !         !         !         !         !         !         !
-1.28E+01 !         !         !  YY     !         !         !         !         !         !         !         !
-1.24E+01 !         !         !         !        YY         !         !         !         !         !         !
-1.20E+01 +----+----+----+----+----+----+----+----+----+----+----+----+----+----+----+----+----+----+----+----+
-1.16E+01 !         !         !         !         !         !         !         !         !         !         !
-1.12E+01 !         !         !         !         !         !         !         !         !         !         !
-1.08E+01 !         !         !         !         !         !         !         !         !         !         !
-1.04E+01 !         !        Y!         !         !   Y     !         !         !         !         !         !
-1.00E+01 !         !         !         !         !         !         !         !         !         !         !
-9.60E+00 !         !         !         !         !         !         !         !         !         !         !
-9.20E+00 !         !         !         !         !         !         !         !         !         !         !
-8.80E+00 !         !         !         !         !       Y !         !         !         !         !         !
-8.40E+00 !         !         !         !         !         !         !         !         !         !         !
-8.00E+00 +----+----+----+----+----+----+----+----+----+----+----+----+----+----+----+----+----+----+----+----+
-7.60E+00 !         !    YY   !         !         !         !         !         !         !         !         !
-7.20E+00 !         !         !         !         !         !Y        !         !         !         !         !
-6.80E+00 !         !         !         !         !         !         !         !         !         !         !
-6.40E+00 !         !         !         !         !         !         !         !         !         !         !
-6.00E+00 !         !         !         !         !         !  YY     !         !         !         !         !
-5.60E+00 !         !         !         !         !         !         !         !         !         !         !
-5.20E+00 !         !         !         !         !         !         !         !         !         !         !
-4.80E+00 !         !  Y      !         !         !         !    YY   !         !         !         !         !
-4.40E+00 !         !         !         !         !         !         !         !         !         !         !
-4.00E+00 +----+----+----+----+----+----+----+----+----+----+----+-Y--+----+----+----+----+----+----+----+----+
-3.60E+00 !         !         !         !         !         !        Y!         !         !         !         !
-3.20E+00 !         !         !         !         !         !         Y         !         !         !         !
-2.80E+00 !         !YY       !         !         !         !         YY        !         !         !         !
-2.40E+00 !         !         !         !         !         !         !YY       !         !         !         !
-2.00E+00 !         !         !         !         !         !         ! YYY     !         !         !         !
-1.60E+00 !         !Y        !         !         !         !         !  YYY    !         !         !         !
-1.20E+00 !         !         !         !         !         !         !    YYY  !         !         !         !
-8.00E-01 !         Y         !         !         !         !         !     YYYYY         !         !         !
-4.00E-01 !         Y         !         !         !         !         !        YYYYYYYYYY !         !         !
 0.00E+00 +----+----YYYYYYYYYYYYYYYYYY--+----+----+----+----+----+----+----+----+-YYYYYYYYYYYYYYYYYYY----+----+
 4.00E-01 !         !         ! YYYYYYYYYY        !         !         !         !         !         Y         !
 8.00E-01 !         !         !         YYYYY     !         !         !         !         !         Y         !
 1.20E+00 !         !         !         ! YYY     !         !         !         !         !         Y         !
 1.60E+00 !         !         !         !    YY   !         !         !         !         !         !         !
 2.00E+00 !         !         !         !      YY !         !         !         !         !        Y!         !
 2.40E+00 !         !         !         !       YY!         !         !         !         !         !         !
 2.80E+00 !         !         !         !         Y         !         !         !         !         !         !
 3.20E+00 !         !         !         !         !Y        !         !         !         !       Y !         !
 3.60E+00 !         !         !         !         !         !         !         !         !         !         !
 4.00E+00 +----+----+----+----+----+----+----+----+-YY-+----+----+----+----+----+----+----+----+----+----+----+
 4.40E+00 !         !         !         !         !         !         !         !         !         !         !
 4.80E+00 !         !         !         !         !   Y     !         !         !         !         !         !
 5.20E+00 !         !         !         !         !         !         !         !         !     YY  !         !
 5.60E+00 !         !         !         !         !     Y   !         !         !         !     YY  !         !
 6.00E+00 !         !         !         !         !         !         !         !         !         !         !
 6.40E+00 !         !         !         !         !         !         !         !         !         !         !
 6.80E+00 !         !         !         !         !       YY!         !         !         !         !         !
 7.20E+00 !         !         !         !         !         !         !         !         !         !         !
 7.60E+00 !         !         !         !         !         !         !         !         !         !         !
 8.00E+00 +----+----+----+----+----+----+----+----+----+----+----+----+----+----+----+----+---Y+----+----+----+
 8.40E+00 !         !         !         !         !         !YY       !         !         !   Y     !         !
 8.80E+00 !         !         !         !         !         !         !         !         !         !         !
 9.20E+00 !         !         !         !         !         !         !         !         !         !         !
 9.60E+00 !         !         !         !         !         !         !         !         !         !         !
 1.00E+01 !         !         !         !         !         !    Y    !         !         !         !         !
 1.04E+01 !         !         !         !         !         !         !         !         !         !         !
 1.08E+01 !         !         !         !         !         !         !         !         !         !         !
 1.12E+01 !         !         !         !         !         !         !         !         Y         !         !
 1.16E+01 !         !         !         !         !         !         !         !         !         !         !
 1.20E+01 +----+----+----+----+----+----+----+----+----+----+----+---Y+----+----+----+----+----+----+----+----+
 1.24E+01 !         !         !         !         !         !         !         !         !         !         !
 1.28E+01 !         !         !         !         !         !         !         !         !         !         !
 1.32E+01 !         !         !         !         !         !         !   Y     !    Y    !         !         !
 1.36E+01 !         !         !         !         !         !         !   Y     !    Y    !         !         !
 1.40E+01 !         !         !         !         !         !         !         Y         !         !         !
 1.44E+01 !         !         !         !         !         !         !         Y         !         !         !
 1.48E+01 !         !         !         !         !         !         !         !         !         !         !
 1.52E+01 !         !         !         !         !         !         !         !         !         !         !
 1.56E+01 !         !         !         !         !         !         !         !         !         !         !
 1.60E+01 +----+----+----+----+----+----+----+----+----+----+----+----+----+----+----+----+----+----+----+----+
```

>> DY <<

```
S I L   S I M U L A T I O N   R E S U L T S
      T                 Y                  DY
0.00000E+0000   2.00000E+0000   0.00000E+0000
3.63636E-0001   1.97780E+0000  -6.77912E-0002
7.27273E-0001   1.95280E+0000  -6.92744E-0002
1.09091E+0000   1.92720E+0000  -7.08166E-0002
1.45455E+0000   1.90107E+0000  -7.25506E-0002
1.81818E+0000   1.87423E+0000  -7.43825E-0002
2.18182E+0000   1.84675E+0000  -7.63701E-0002
2.54545E+0000   1.81848E+0000  -7.85632E-0002
2.90909E+0000   1.78937E+0000  -8.09676E-0002
3.27273E+0000   1.75928E+0000  -8.35974E-0002
3.63636E+0000   1.72820E+0000  -8.65753E-0002
4.00000E+0000   1.69600E+0000  -8.98708E-0002
4.36364E+0000   1.66242E+0000  -9.36530E-0002
4.72727E+0000   1.62735E+0000  -9.79967E-0002
5.09091E+0000   1.59065E+0000  -1.03007E-0001
5.45455E+0000   1.55192E+0000  -1.08969E-0001
5.81818E+0000   1.51082E+0000  -1.16161E-0001
6.18182E+0000   1.46676E+0000  -1.25072E-0001
6.54545E+0000   1.41897E+0000  -1.36524E-0001
6.90909E+0000   1.36639E+0000  -1.51896E-0001
7.27273E+0000   1.30739E+0000  -1.73915E-0001
7.63636E+0000   1.23830E+0000  -2.08829E-0001
8.00000E+0000   1.15186E+0000  -2.73977E-0001
8.36364E+0000   1.02771E+0000  -4.38281E-0001
8.72727E+0000   7.66837E-0001  -1.26763E+0000
9.09091E+0000  -1.43093E+0000  -1.20121E+0001
9.45455E+0000  -2.00552E+0000   6.58190E-0002
9.81818E+0000  -1.98116E+0000   6.75983E-0002
1.01818E+0001  -1.95624E+0000   6.90709E-0002
1.05455E+0001  -1.93077E+0000   7.06449E-0002
1.09091E+0001  -1.90471E+0000   7.23051E-0002
1.12727E+0001  -1.87799E+0000   7.41306E-0002
1.16364E+0001  -1.85056E+0000   7.61286E-0002
1.20000E+0001  -1.82238E+0000   7.82527E-0002
1.23636E+0001  -1.79340E+0000   8.06346E-0002
1.27273E+0001  -1.76349E+0000   8.32411E-0002
1.30909E+0001  -1.73259E+0000   8.61328E-0002
1.34545E+0001  -1.70053E+0000   8.93853E-0002
1.38182E+0001  -1.66713E+0000   9.30659E-0002
1.41818E+0001  -1.63237E+0000   9.73302E-0002
1.45455E+0001  -1.59587E+0000   1.02255E-0001
1.49091E+0001  -1.55746E+0000   1.08074E-0001
1.52727E+0001  -1.51669E+0000   1.15043E-0001
1.56364E+0001  -1.47302E+0000   1.23717E-0001
1.60000E+0001  -1.42577E+0000   1.34742E-0001
1.63636E+0001  -1.37371E+0000   1.49530E-0001
1.67273E+0001  -1.31507E+0000   1.70768E-0001
1.70909E+0001  -1.24710E+0000   2.03814E-0001
1.74545E+0001  -1.16323E+0000   2.63587E-0001
1.78182E+0001  -1.04560E+0000   4.07711E-0001
1.81818E+0001  -8.15688E-0001   1.05136E+0000
1.85455E+0001   8.55434E-0001   1.39859E+0001
1.89091E+0001   2.00824E+0000  -6.46404E-0002
1.92727E+0001   1.98400E+0000  -6.74371E-0002
1.96364E+0001   1.95913E+0000  -6.88947E-0002
2.00000E+0001   1.93367E+0000  -7.04272E-0002
 91.23   SECONDS in execution
 106.0   KBytes left in Long Heap memory
```

Figure 3.2: The VANPOL.LST file (output example).

You can of course run any of the models in the '*.SIL' files in a similar manner. These models have been chosen in order to demonstrate most of the facillities of the SIL language. We will return to some of them later.

You should notice, the two different output statements. WRITE produces tables of the requested variables whereas PRINT produces a line printer plot. For the PRINT statement you can have either a simple list of argument variables like (Y, DY) or you can have an argument of the form (Y(DY)) . In the first case μSIL will produce a plot of the variables versus model time; in the latter case you get a phase plane plot; variable Y is plotted versus variable DY .

3.1. Running the μSIL System.

You have now run μSIL for the first time; we shall now look at how things are organized and see some of the options in the system.

The μSIL system is activated by typing (at the DOS prompt) the command SIL followed by the name of the model. If you omit the model name μSIL will prompt you for it by a line starting with MODEL :_ . You can then enter the command line parameter (the name) as before.

The name of the model is actually the name of the file in which the model is stored and μSIL allows you to enter file names with their (legal) DOS path. If you have made the SIL directory on your C-disk (usually the harddisk) then the previous example could as well have been run (from any directory) with the command: SIL C:\SIL\VANPOL . If μSIL cannot find the <name>.SIL file because you either mistyped it or it is in another directory you will get an error message on the screen saying which file could not be found.

If the file <name>.SIL is present, then μSIL will after a short initialization start reading it. At the same time it will echo (write) all that is read to the file <name>.LST . If the model is not correct according to the SIL language definition all error messages and additional information (a line number indicating the position of the error etc.) will also be printed in that file. In this situation, the DOS prompt will appear immediately after the model has been read; there will be no attempt to run the model unless it has succesfully passed the initial model control and analysis (compilation).

When μSIL is actually solving the problem (indicated by a screen line giving the step number and the model time) you may interrupt the computations by pressing either the 'S' or the 's' key. This will, after a short while, force μSIL to stop the simulation and generate output in the <name>.LST file as if the end of the integration interval has been reached. Normally, the Break-key is deactivated by the μSIL system, so it will not work at all. If for some reasons an emergency STOP is required only the standard reset (Ctrl-Alt-Delete) will work and all results produced so far will be lost. The reason for this is that during the solution process the μSIL system stores the output values in RAM memory; when the solution process is finished, the output data is processed according to the specifications in the model.

Another way of stopping the solution phase of a model is to use some of the built-in control options in the SIL language. First of all, there exists a predeclared variable called MAXCPU, and giving it a non-zero value will cause μSIL to check regularly the computing time used (in seconds) with the bound prescribed and stop the solution of the model if the bound is exceeded. Also, you may for each of the variables in the model specify a 'validity-range' (see section 6.2.1.3) and any variable exceeding its range will cause the solution process to be stopped. This way of stopping the solution process is useful because all the results produced so far are printed.

3.2. Your First SIL Model.

We have in the previous chapter given you the SIL model for the Van der Pol equation. Written mathematically the equation looks like this

$$y'' = c\,(1 - y^2)\,y' - y$$
$$t \in [0,20], \quad y(0) = 2.0, \quad y'(0) = 0.0, \quad c = 10.0\ .$$

Comparing this with the VANPOL.SIL file you can see how the various mathematical terms are interpreted in the SIL language.

Now, it is your turn to solve a problem. This one will be easy. Make a SIL model of the following mathematical problem,

$$y'' = -\,y$$
$$t \in [0,6.5], \quad y(0) = 0.0, \quad y'(0) = 1.0\ .$$

By inspection you can verify that y = SIN(t) is a solution. Differentiating y once with respect to t gives y' = COS(t) and differentiating twice gives y" = -SIN(t) = -y . This solution agrees with the initial conditions. In order to check the accuracy of the μSIL simulation procedure you can compute the errors $e_1 = y - \sin(t)$ and $e_2 = y' - \cos(t)$ and use WRITE to output them.

An easy way to solve the problem is first to copy VANPOL.SIL to MYPROB.SIL using the DOS COPY command. Second, take your favorite editor and modify the file MYPROB.SIL so that it corresponds to the problem stated. This should be easy, specially if you notice the similarities between this problem and the Van der Pol equation.

In order to get the error of the numerical solution process you can just insert the statement WRITE(Y-SIN(T), DY-COS(T)) and separate it from the next statement with a ';' (semicolon). Alternatively, you can declare two extra variables, E1 and E2 say; assign them the values Y-SIN(T) and DY-COS(T) respectively, and then WRITE them out.

Save your model MYPROB in the file MYPROB.SIL and run μSIL on it with the command SIL MYPROB . If you have made some errors, look at the file MYPROB.LST to find out what the errors are. Generally, when μSIL locates an error at a particular position, the reason for the error can be found somewhat before that position (sometimes on the preceding line). When your model is without errors and is correctly executed, you can find the results in the file MYPROB.LST .

If you look at the table of the errors you will notice that they are approximately 10 times larger than the values specified as error tolerance (ABSERROR, RELERROR). This observation need not always hold but one normally observes that the true error (the global error) is larger than the error tolerance bound specified. The bound is applied locally; that is, on each step of the integration. The global error includes the effect of accumulating all local errors and is therefore normally greater than the bound specified on the local error.

The variable T is used as model time; usually model time corresponds to physical time but it it not a necessity. Model time always represents the independent variable; that is, the variable with respect to which differentiation in a dynamic model is performed. Only one model time is allowed in a model and it will implicitly appear in all output statements.

With your first correct SIL model you can proceed to the next chapter and learn how to use the μSIL graphics.

3.2.1. Using the Graphic Command System.

Assume that your model MYPROB.SIL is correct; that is, it can be compiled correctly and produces the correct results. You can now call the μSIL system with the following command

SIL MYPROB (G

to activate the graphic display system. Of course, you can also use the (G option if your model is not correct and you still get the <name>.LST file and no execution. If your model is in order, after a short while (μSIL must compile the model again) your screen should look something like fig. 3.3. You are now in the graphic command system. μSIL attempts to detect automatically the kind of graphics card on your system; if your screen does not look at all like fig. 3.3, μSIL has propably detected a wrong type of graphics card; chapter 7.3 will show you how to solve this problem.

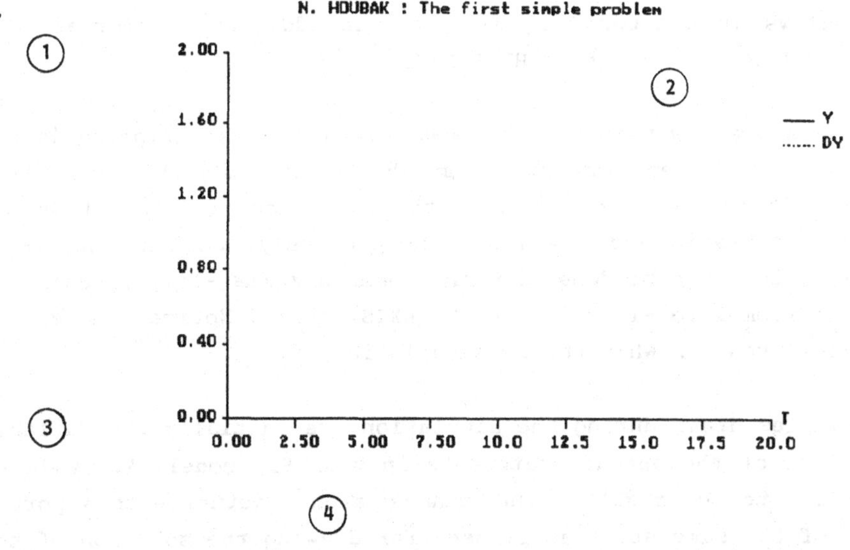

Figure 3.3: Initial graphic screen layout.

The screen is divided into 4 fields numbered (1) to (4) on the figure. The large graphic field ((2)) is for μSIL to display the solutions; you will notice that the axes are already drawn and scaled. The scaling is based on the initial values of your variables but you can easily change it, see section 7.4. To the left of the graphic display area you will find the command input field ((1)); when you type in commands they will be echoed in this field. Below the graphic field and across the screen you will find the output field ((4)); this is used by μSIL to show values of variables and to display parts of the SIL.HLP file when help is needed. Between the input field and the output field there is a small (3 lines) status field ((3)); here μSIL will write the running step number, the actual model time and perhaps an error code during the simulation.

You should notice, that no solution of the model has been performed yet. μSIL has stopped and by the prompt ' > ' in the input field it indicates that it is waiting for you to enter a command. The most commonly used commands are for rescaling the axes, defining new plotting variables, displaying/changing values of variables, performing simulation, and getting help. The command HELP (in either lower or upper case letters) is very useful; when activated it will display in the output field the list of all the graphics subsystem commands. We shall shortly describe some of these now; a more detailed description follows in chapter 7.3. If you want additional information on the command XAXIS, just type HELP XAXIS .

μSIL always draws within the area edged by the axes; clipping is performed outside. There are two commands for scaling the axes, XAXIS and YAXIS; they are identical with the only exception that they operate on the X-axis and the Y-axis respectively. Scaling the X-axis to from -1 to 1 can be done with the command XAXIS(-1:1) . Making the Y-axis go from 1 to -1 is done with YAXIS(1:-1) . Notice, the axis is immediately redrawn when the command is issued.

The variables drawn during the simulation are initially the variables in the last of the output statements in your SIL model. You will see their names to the right of the drawing area together with a portion of line of the same style as is used for drawing the solution of that particular variable. Only variables in output statements (WRITE, PRINT or PLOT) are sampled for output during a simulation; you can add (temporarily) one extra output statement by the PLOT command, like PLOT(Y(DY)), provided you have declared the variables Y and DY in your model. This of course becomes the last output statement and the variables defined in it will be the variables drawn during the simulation. After issuing this command you will see the names to the right of the drawing area are changed to the name Y (the variable DY is not plotted).

Had the drawing area been scaled correctly and had the last output statement contained the variables to be drawn during the simulation none of the above commands need to have been issued and you could immediately issue the command SIM to start the solution process. If the integration interval of interest is different from the interval given in the SIL model, you may used the command SIM(0:12.3) in order to solve the problem in another time interval.

During the solution process you will notice that the actual step number and the actual model time are printed in the status field; during the solution process the variables from the last output statement are always simultaneously drawn in the drawing area. You may interrupt the solution process by pressing the 'S'-key or the 's'-key. When the solution has finished you will be notified by a 'beep' and the prompt ' >' will appear on the next line in the command input field.

Suppose you want to see the behaviour of the variables E1 and E2 (the 2 variables for holding the errors in the example); this can be done by the command DRAW(E1, E2) . Due to the scaling of the drawing area this becomes a straight line. Rescale the Y-axis by the command YAXIS(-1E-5:1E-5) (E is an exponentiation mark, "'" could have been used as well) and the X-axis by the command XAXIS(0:10); make a drawing with the command DRAW(E1,E2) ; you can now see the errors. The DRAW command allows you to plot any sampled variable versus time or versus any other sampled variable (a variable is sampled when it appears at least once in an output statement) without performing any simulation. It is essential, that the variables drawn have been sampled during the most resent simulation. μSIL will do any interpolation of the sampled values necessary, if the DRAW command involves solutions that are not sampled at the same positions in time.

The command END terminates the graphics command system and leaves you at the DOS command prompt. You can also leave the graphics command system by the command EDIT . This command invokes the μSIL line editor on the model file. This will be explained later.

Variables plotted during the solution process (the SIM command) will normally appear as very smooth solutions. Both output points and step points are used for drawing the curves. The DRAW command only uses the output points and may therefore on some problems produce a solution curve which is not quite as good as the one produced during a SIM command. A more frequent sampling of the variable may solve that problem.

3.3. A Stiff Complex Problem.

In this section we will describe the consequences of stiffness and what is stiffness. Actually, it is very difficult to give a precise mathematical definition of stiffness, so we will mainly consider its consequences.

As an example, take the Van der Pol equation from before; change the value of the parameter EPS from 10 to 100 and the time interval from [0,20] to [0,180]. If you try to solve the problem again, the computing time will be of the order of half an hour or so. Looking at the statistics printed after the model listing in the VANPOL.LST file it appears that the integration procedure has used 20967 steps now instead of the 400 used for solving the original problem; the number of function calls is not printed because it exceeds 32767 (2^{15}-1). The number of divergences detected in the iteration process is also enormous, indicating that something is very wrong.

The solution will have almost the same shape in both cases (neglecting the different time scale). The cause for the increase in computing time is the increased number of steps. Suppose you could plot the accumulated number of steps versus time, you would see that a lot of steps are used where the solution changes dramatically; that is expected and acceptable. In the time intervals where the solution is almost constant an even larger number of steps are used because this part spans most of the time scale. The method does not recognize that the solution is almost constant; it sees indefinitely many steep transients going into the (constant) solution and they cause the integrator to take small steps due to its numerical stability.

The problem of stiffness in the Van der Pol equation becomes worse the higher the value of EPS . You cannot tell whether a particular problem is stiff or not on before hand, that property depends on the required accuracy of the solution, the time interval, to some extent the numerical method used, and of course on the system of differential equations itself.

The integrator in μSIL has a special option which is specially well suited for solving stiff problems efficiently. A predeclared variable (METHOD) has the default value 119; if that value is changed to 139 by a statement of the form

```
METHOD := 139;
```

in the Van der Pol model, μSIL will use its stiff option for solving the problem. Predeclared variables are described in section 6.2.2.3. Now run the (stiff) Van der Pol problem again with the stiff option used for the integrator and notice that the computing time now is one and a half minute and the number of steps is 325 and only 2467 function calls are used. Normally, the computational work is increased if the stiff option is used for solving a non-stiff problem; therefore, the non-stiff option is the default option. Since many simulation problems are stiff, though, it could be worth while to try the stiff option in order to reduce the overall computing time.

Stiffness has many aspects; we cannot cover them all, but section 5 contains a more detailed (but not complete) description of the problem.

3.3.1. Integrator, System Variables, Debugging.

The integrator used in μSIL is a standard integrator called STRIDE written by Burrage, Butcher, and Chipman (1980). You don't have to worry about how your model is solved and the use of STRIDE also ensures that you don't need to worry if the solutions are correct; they are correct to within the accuracy specified. We could have used any standard (PASCAL written) integration procedure, but STRIDE was preferred because of its reliability and robustness. Of course, you can get incorrect results from μSIL; if you specify an absolute error tolerance of 10^{-3} (say) and one of your variables is 10^{-4} (say) in magnitude; this variable being wrong can totally change the solution.

μSIL is prepared for having several integration routines but only 1 is implemented at the moment. The predefined variable INTEGRATOR is used to select between those currently available. INTEGRATOR is an example of a predefined variable in the language; the complete list can be found in section 6.2.2.3. Their default values can be changed by statements of the form

<id> := <number>

where <id> is any of the predefined variables and <number> is either an integer or a real number. The most commonly used predefined variables are

ABSERROR for setting the absolute local error tolerance to something that differs from the default 10^{-5}.

RELERROR for setting the relative local error tolerance to something that differs from the default 10^{-5}.

MAXCPU makes μSIL terminate integration if it has been running for too long (in seconds). The default is no limit on time.

METHOD is used to change between different options in the integrator. In STRIDE it is used to switch between a stiff (139) and a (default) non-stiff (119) integration process.

MAXSTEPSIZE limits the maximum time stepsize taken by the integrator. The default is no limit on the stepsize. Output points are interpolated; they do not impose a limit to the stepsize.

The values assigned to MAXCPU and METHOD must be integers whereas the other predeclared variables accept reals. Statements that redefine the predefined variables can be anywhere in a SIL program with the exception that they should not appear in an IF THEN ELSE construction. If a variable is changed several times only the last one will have effect. Most of the predeclared variables are also accessible in the graphic command system. Since the variables are predeclared and for internal use, they should never be used as 'model variables'.

Primarily for internal use, there is a built-in facility for debugging. A line in the source program like

$DEBUG,2

with $ in column 1 and DEBUG in capital letters will bring the μSIL runtime system into a debugging mode. In this mode μSIL will, during the integration, write the solution after each completed step in the .LST file. DEBUG,3 will in addition to what DEBUG,2 writes also print the status of the model when it is changed by a discontinuity. DEBUG,4 will during the integration write a line in the .LST file each time a variable is updated giving the new value of the variable. DEBUG numbers from 5 to 9 are primarily intended for debugging the compiler and should normally not be used since they will cause a large amount of output to be generated during the compile phase.

Should you discover a strange behaviour of the μSIL system which possibly could be the result of an error in the system, run the smallest possible model that illustrates the error for the shortest possible time with a $DEBUG,9 in the very first line. Send your model file and the resulting .LST file on a diskette to the author; don't print the .LST file, it will normally be extremely large!

There are other $ commands available in the system, they are

$NOLIST suppresses the listing of the SIL model in the .LST file from this line.

$LIST resumes the listing of the model. When macros are called, the expansion will be printed in the .LST file if the $LIST is in the level of 2.

$TITLE= will generate a formfeed in the .LST file and set a new running title on the page header. The last title will be used initially in the graphic command system as plot title.

It is essential that the characters following the $ sign are as shown, capitalized. These control lines will not appear in the listing of the model.

3.3.2. The Editor.

The μSIL system has a built-in syntax checking interactive line-oriented command driven editor. It can be activated either when calling μSIL or from the graphic command system. Calling μSIL with the command

```
SIL <name> (I
```

will automatically invoke the μSIL editor on the model specified by <name>. From the graphic command system the editor can be activated on the actual model by the command EDIT . In both cases the editor will type on the screen

```
SIL, Interactive Editor
HELP is available

*** TOP of file
=> _
```

The prompt => indicates that the editor is ready for a command. You can get a list of all the available commands by typing HELP . We will shortly describe the most commonly used commands. First though, we will describe the principles of the editor and how it works.

The editor is a line editor, and it is not suited for typing in large models; use it in connection with minor changes in models, adding equations, replacing expressions, or increasing the number of output statements. The editor always holds a copy of the model (in text format) in core; this storage reduces the available amount of storage for the model and the sampled solution. For very large models there may not be storage enough for the editor to work.

A line editor, such as that in μSIL, always has a current line. This line is normally displayed immediately above the editor prompt. The editor has a virtual line before the first line (called TOP of the file) and after the last line (called BOTTOM of the file). You can make either of these virtual lines the current line by the commands TOP and BOTTOM, respectively. Normally, the editor is in COMPILE mode; this means that the model from the top line to and including the current line has been checked by the μSIL syntax analyzer. When the prompt appears after the BOTTOM command, the whole model is analyzed.

The most important editor commands are shortly described below. You can type any of these commands as response to the editor prompt => . The editor, though, allows you to type several commands on the same line; they must be separated with a ';' (semicolon). You can get a complete description of any of the commands on-line by typing HELP followed by the command that for which help is wanted. In this chapter all the commands are written in upper case; this is for clarity: in μSIL it is not required. Commands can be abbreviated down to one character and commands need not be spelled correctly; some simple recovery from typing errors is attempted by the editor on 'non-destructive' commands.

One of the most important commands is the one used for picking another line as current line. Positioning can be done by giving relative line movements, +5 moves five lines ahead, and -10 moves ten lines backward. When the current line is printed above the editor prompt, the linenumber relative to the virtual top line is also printed. When moving 10 lines ahead, say, you can get all the lines listed on the screen with the command LIST +10 (for short L+10).

A new line (or several lines) can be generated and inserted just after the current line with the command INPUT. In input mode you will have the next linenumber prompting you for the contents of the next line. An empty line (not a blank line) terminates the input mode. As the lines are entered and terminated by <enter>, the syntax is analyzed by μSIL. An error will terminate input mode. The command I4 automatically terminates input mode after 4 lines of input. The line numbers of the lines following these new lines are automatically adjusted. The last line inserted will become the current line.

The current line is permanently removed from the model with the command DELETE . The next line becomes the current line. The command DEL-1 deletes the current line and makes the preceding line the current line.

The FIND command is case sensitive; the string to search for is only located if it matches correctly with upper and lower case letters. The command FIND /abc/ will not locate the string Abc . The FIND command starts searching in the line following the current line. The first line with a match is typed on the screen and becomes the current line. FI7abc7 searches for the string abc as does the command F,abc . The command FIND alone searches for the string from the most recent preceding FIND or REPLACE command.

REPLACE is used for replacing a string of text in the current line with another string. REPLACE /abc/ABC/ will replace the (lower case) string abc with the (upper case) string ABC . The first non-blank character following the REPLACE command is assumed also to be the delimiter between the two strings. After a REPLACE command FIND without a string will search for the first string and REPLACE (again and without parameters) will repeat the replacement on the current line. The REPLACE command, if succesfull, will cause μSIL to recompile the model from the top until and including the current line. Some disk activity will normally be seen during this process.

The editing is terminated by the END command. The model will be compiled to the bottom; if it is correct, it is also written back to the file it was read from, and the μSIL system will go into the graphic command system. The μSIL editor will not save a model which has syntax errors. The command ABORT terminates the editor without saving the model and without going into the graphic command system.

4. Building Models.

The μSIL system and the SIL language are specially designed to make it easy to do simulations of systems. In this chapter we will look closer on what we understand by simulation and what models are. We will particularly define the concepts necessary to fully understand the SIL language and the types of models it can be used for.

There are three primary reasons for doing simulations. Number one is money, number two is time, and the third reason is that for some systems testing the system means changing the system. Some systems are so big (a bridge), so expensive (a space shuttle), or of such a short duration of time (some chemical reactions) that the only economical way to get some knowledge about the system behaviour is to do simulations. The effect of interest may be so long lasting (a reliability test) that the product is out-moded if a true time scale test should be performed. In many political, economical, and ecological systems you cannot make an experiment with the system without also changing the system; this makes it impossible to reproduce realistic results.

There are many ways of doing simulations. An obvious one is to build a scale model of the system and test its behaviour to some known laboratory conditions. This can cut the costs for the experiments dramatically and the results can be as good as if it were the real system that had been in the laboratory. Another way of simulating a system is to make some kind of analogy or equivalence; the system is assumed to behave like another system (because their governing equations are similar). The second system is much cheaper to build and measure. Very often electrical circuits have been used as analoges. The third way of doing simulations is by means of a computer. The model built is mathematical equations describing the system; they are solved on a computer and the solutions are assumed to describe the behaviour of the physical system. μSIL can make the last way of simulating a system very easy.

Building mathematical models of a system is not an easy task. Very often, building the model is the main problem whereas solving the model can be done by standard techniques. Actually, all the problems you can solve using SIL, you can also solve by making a program yourself; many of the facilities in SIL are implemented using standard techniques. On the other hand, you can describe and solve the problem much easier and faster using SIL, and you need not worry very much about programming errors, numerical properties of the solution method, consistency of the problem, and the generation of output. This is done once and for all in the μSIL system.

The quality of a mathematical model has nothing to do with the size of the model (the number of equations). The important point is that the model should describe the behaviour of the physical system to within the accuracy desired. As an example take Ohms' law; most resistors obey this law except when red-hot or when exposed to high frequency current. In these two cases Ohms' law is inaccurate. It should be emphasized that all parts of a model should be accurate to within almost the same accuracy. There is no point in having a model given as (say) 100 equations when 5 well chosen equations are enough to give the reqired final accuracy. Unfortunately, this is often what happens when large models evolve.

Answers to many questions don't need large models. Sometimes answers from small models may be even better than those from the larger models because of a minor influence of rounding errors. 'Small model results' are certainly much cheaper to produce; small models can be run more often under different conditions and this gives a much better insight in and understanding of the system. In conclusion, one may say that **results are not better just because they have required a lot of computations.**

Keep models simple. When building a model, start with the most simple version you can think of. It should of course contain the physics that is to be modelled but nothing more. As an example take a heat-exchanger as a component in a larger system. It can be modelled using partial differential equations describing in details the heat transfer process from one fluid to another giving also the internal temperature distribution. In order to solve this to within a reasonably accuracy some hundred equations might be required. If the heat exchanger is part of a larger network with several other heat exchangers the total model will be enormous. In this big model a lot of work is spent on finding the internal temperature distribution of all the heat exchangers; if that is of no interest a lot of work is wasted.

The fine model of the heat exchanger could (should) however be used to determine a simple model (1 or 2 equations) for how the heat exchanger behaves as a component in a network. This simple model can be verified and calibrated by experiments and used in the model of the network. With simple models in the network, it is much easier to analyze, it is easier to solve, and it is easier to get insight into its general behaviour.

4.1. Phases of a Simulation.

It is very difficult in general to say how to build good models and make them work. This requires more than a textbook. In this section we will shortly describe the phases one has to go through when building models and running them. Let us start from the very beginning.

We define our system to be a little piece of the real world. This is our physical system. We are interested in knowing how this system responds to some conditions imposed on the system from the outside world. This is what our simulation is going to tell us. By the way, one condition imposed on the system could be suddenly to isolate it from the outside world.

The very first thing to do is to get a 'picture' of the system; this we call a physical model. Usually it is a system diagram, a process sketch, or a network structure diagram. When making this physical model it is very important to include all components and effects that must be taken into account and to exclude all effects and components that are to be neglected. As an example of this we can mention that a resistor normally obeys Ohms' law $V = R\ I$. However, if the current through the resistor is high frequency the physical component (the resistor) may actually act as another electrical component (a capacitor, an inductor or as a network of simple components).

The physical model is now a basis for building a mathematical model of the system. Each of the components/effects in the physical model will have some kind of governing equation(s). Some of these are basic natural laws, Ohms' law being an example. Combining these equations with the basic balance equations (what goes in minus what goes out is what remains in) gives the set of mathematical equations that determine the behaviour of our model. We call it our mathematical model.

Mathematical models can all be divided into different groups depending on the types of equations that appear. The first group of models consists of algebraic equations; those models we call static models. Models in the second group all contain at least one ordinary differential equation (ODE) which can be solved as an Initial Value Problem (IVP); we call them dynamic models. The third group is for models mainly consisting of partial differential equations (PDE); in many cases these problems must be solved as Boundary Value Problems (BVP). It is very difficult in general to treat models from the third group, and μSIL was never meant for solving PDEs. Problems from the remaining two groups can in general be described as a system of first or higher order ordinary differential equations coupled with a system of algebraic equations; and this is what μSIL is built to solve. Further, in μSIL all models, linear and non-linear, are treated as non-linear models.

The mathematical model can be solved either analytically or by introducing a numerical model. The differential equations are discreticed using some kind of discretization algorithm (a numerical method). The algebraic equations are solved by using an iterative method. Further, numerical techniques are implemented in order to make the whole solution process work efficiently (detecting discontinuities, interpolating solution for output points, etc.).

Implementing and running the numerical model on the computer produces the results. These results are often claimed to be the way that the system behaves without any concern to the errors made in each step of our simulation. The most important part of our simulation is therefore model verification. We must make sure that the results produced are good estimates on the behaviour of the physical system. This is sometimes possible by measurements on the system or comparison with historical data; in other cases it can be quite impossible to verify the model. Schematically, the modelling process is given in fig. 4.1. The lines backward from the verification stage indicates that this stage may lead to modifications everywhere in the model.

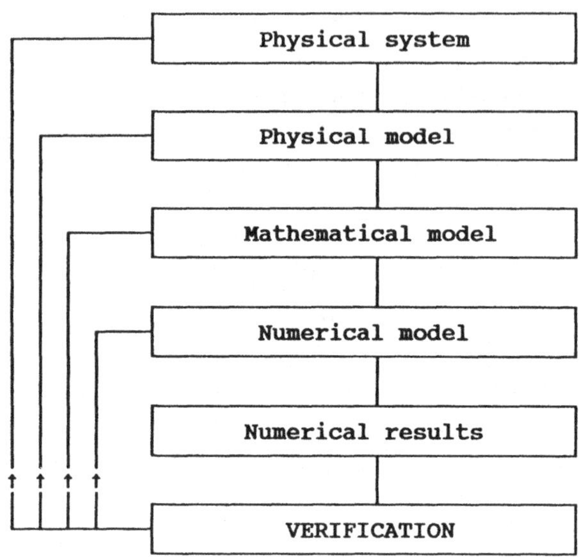

Figure 4.1: The different stages of the simulation/modelling process.

The μSIL system is very useful when passing from the stage 'mathematical model' to the stage 'numerical results'; but it relies very heavily on the mathematical model being in agreement with the physical system. To some extent the strong syntax of the SIL language will prevent inconsistent models and thereby lead to better models.

The μSIL system has facilities that are helpful in the verification of a model; more specifically, in the graphic command system, the user may interactively change initial values as well as parameter values and get the different solutions on the same plot. This can give a good idea of the sensitivity of the model to either the initial values or the values of some of the parameters.

4.2. Definitions.

Since μSIL has as its basis mathematical models, we will now more closely describe these and make definitions used in the rest of this manual.

PARAMETER: Some physical quantities are constant either by definition or by assumption (the speed of light, pi, dimensions, ambient temperature, mass, spring constant etc.); such a quantity is called a **parameter** in SIL. During a simulation (solution) the value of a parameter is never changed.

VARIABLE: The most central role in any mathematical model is played by variables. The value of a variable represents a physical quantity; normally, a quantity which can be measured and often a quantity which changes in time.

Definition 1: The value of a variable is always determined by some kind of equation; the number of equations of a certain kind must match the number of corresponding variables.

We can divide the equations into three main groups; **algebraic** equations, **differential** equations, and **difference** equations. The following definition is based on considering only ordinary differential equations and hence only differentiation with respect to one independent variable (usually the time); also, we exclude difference equations and concentrate on continuous systems.

Definition 2: A differential equation is an equation from which a **derivative** is determined; all other equations are algebraic equations. Variables, which are solutions to differential equations, we call **dynamic variables**; all other variables are called **static variables.**

Static variables as well as derivatives can appear in either of two modes, explicit or implicit. In the deciding whether a static variable (or a derivative) is explicit or implicit, we assume the dynamic variables and parameters to be known quantities.

Definition 3: If a static variable V is assigned a value by a statement of the form

V := <expression> ,

then V is said to be a **static explicit variable.** It is required that <expression> neither directly nor indirectly depends on V itself. If it does, V is a **static implicit variable** and an implicit assignment statement of the form

<number> := <expression> ,

must be used to specify the equation (<number> is the absolute bound on the residual error). The same rule applies to the definition of **explicit derivatives** and **implicit derivatives**.

We give the various models names based on the type of variables that appear in them.

Definition 4: A model containing dynamic variables is called a **dynamic** model. A model without dynamic variables is called a **static** model.

A dynamic model must have at least one differential equation and it must therefore have a solution which varies with model time. Static models are normally used for finding steady state solutions (solutions in the limit as model time goes to infinity). Such a solution is static. In two situations we have static models which vary in time. One may couple a static and a dynamic model; during the solution of the dynamic model, the static equations are required to be satisfied; mathematically we call it a differential algebraic problem. The model used for answering the question, "how does the static solution change if we vary one of the parameters in the model?" is a static model; the model time though, can then be used to simulate the variation of the parameter and we call the model quasi static (or quasi stationary).

In simulation it is very common that a part of the model only is active in a part of the solution domain, while another part of the model is active on the rest of the domain. We call the different parts of the model the **states** of the model.

Definition 5: If a model contains more than one state, we call it a **discontinuous** model. Variables cannot be undefined (without a value) in any of the model states.

Difference equations are used for defining discrete systems.

<u>Definition 6</u>: Discrete **variables** are updated with certain time intervals, the **sample time**; between updates they remain constant. Because of the limited number of updates, a discrete variable may depend recursively on itself.

A **discrete** model contains discrete variables; it is also a discontinuous model and it may be combined with either a dynamic model or a static model.

4.3. Simple Models.

In order to illustrate some of the properties mentioned previously we will expand a very simple problem, which has an analytic solution, into a rather complex problem without analytic solutions. During this expansion we will introduce some of the definitions mentioned in the previous chapter.

The very first problem is to solve the cubic equation

$$x^3 = c$$

where c is a constant. This problem is a static problem and it has a unique analytic solution. A SIL program describing and solving this problem explicitly (when c is positive) is given in fig. 4.2.

```
BEGIN
VARIABLE  X;
PARAMETER  C(2.7);   (*  Find the cube root of 2.7   *)

(*  Set up the equation    *)

X := EXP( LOG(C) / 3);

(*  Print the result       *)

WRITE(X)
END.
```

<u>Figure 4.2</u>: A SIL program for computing the cube root.

Changing the parameter C makes it possible to determine the cube root of any positive real number. The variable X is an explicit static variable, the equation for X in fig. 4.2 is an explicit static equation, and the model is a static model. It is also possible to solve the problem without using the analytic solution; that is, make SIL solve the problem numerically by using an implicit static equation. This is done in fig. 4.3.

```
BEGIN
VARIABLE  X(1.2);  (*  Arbitrary initial value  *)
PARAMETER  C(2.7);  (*  Find the cube root of 2.7  *)

(*  Set up the equation   *)

1.0'-6 := X*X*X - C;

(*  Print the result      *)

WRITE(X)
END.
```

Figure 4.3: A SIL program for finding the cube root numerically.

Now there is no requirement that C is positive and this makes it possible to determine the cube root of any real number. The variable X is an implicit static variable, the equation for X in fig. 4.3 is an implicit static equation, and the total model is a static model. It is of course more expensive (in CPU time) to solve the problem implicitly but in many cases it is necessary and convenient.

We can expand the problem to the following: "What does the solution x look like for the equation

$$x^3 = c^2 \tag{4.1}$$

when c varies in the interval [-1, 1]"? This problem also has an analytic solution. It can be solved in several different ways using SIL; fig. 4.4 illustrates one of them. Notice, the problem has been reformulated a little since a differential equation has been added. The idea behind that is: The variation of the 'constant' c is simulated by requiring c to be the solution to the equation

$$c' = 1 , \quad c(0) = -1 , \quad t \in [0,2]$$

and simultaneously requiring equation (4.1) to be satisfied. Since the problem has an analytic solution, it is possible to find the error in the implicit solution.

```
$TITLE=Index 2 problem
BEGIN

VARIABLE      C(-1),    X(0.9),    XERR;
DERIVATIVE A(C);
TIME T(0:2);

A := 1;

(*  The equation  x^3 = c^2
    must be satisfied for varying  c  *)

1.0E-6 := X*X*X - C*C;

(*  Compute the error on  x   *)

XERR := EXP(LOG(C*C)/3) - X;

(*  Print the solution and the error  *)

PRINT(200, X, XERR)

END.
```

Figure 4.4: A SIL program solving $x^3 = c^2$ for $c \in [-1, 1]$.

When modelled this way, we get a DAE (Differential Algebraic Equation) system to solve and our model becomes a dynamic model. The variable C is a dynamic variable, and X is an implicit static variable. The main difficulty in solving this particular problem numerically, is a singularity in the Jacobian matrix of the algebraic equation (4.1) for $c = 0$; we say the problem is locally index 2 (the index problem is described in chapter 5.1.

Examples of output from this model processed by the graphic command system are shown in the figures 4.5 and 4.6. The displayed curves are the solution X and the error of the solution XERR, respectively, plotted versus C .

It is very easy to change the equation (4.1) in such a way that there is no analytic solution; here is an example:

$$x^3 e^x = c^2 .$$

In fig. 4.7 we show a SIL program which solves this problem.

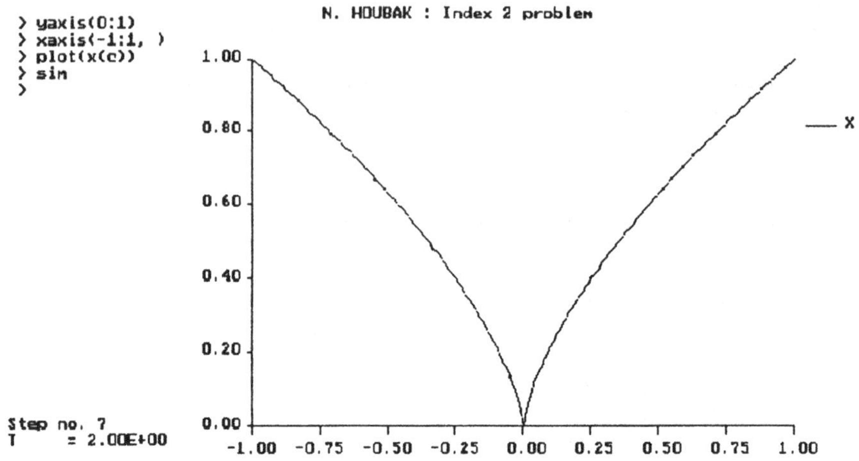

Figure 4.5: The graphic screen dumped onto a printer after solving the problem in fig. 4.4.

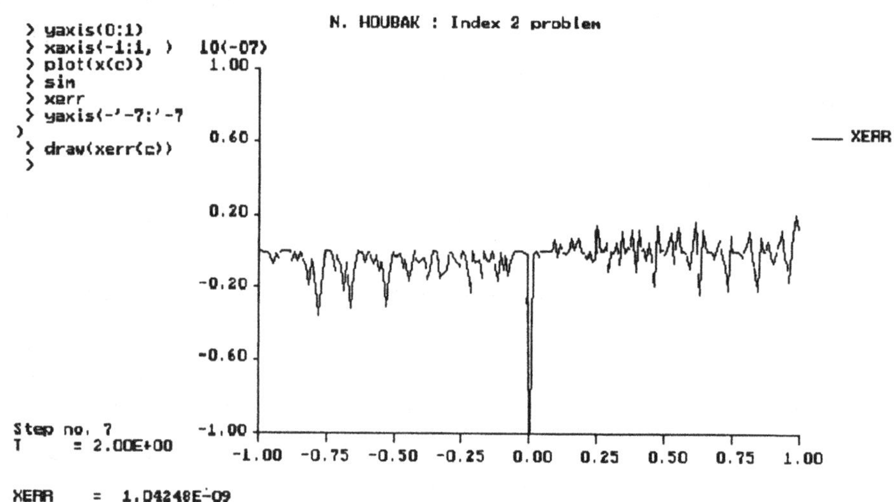

Figure 4.6: The graphic screen dumped onto a printer after having reDRAWn the error on the solution from fig. 4.4.

```
$TITLE=Index 2 problem
BEGIN

VARIABLE     C(-1),    X(0.9);
DERIVATIVE A(C);
TIME T(0:2);

A := 1;

(*  The equation  x^3 e^x = c^2
    must be satisfied for varying  c  *)

1.0E-6 := X*X*X * EXP(X) - C*C;

(*  Print the solution  X  versus  c   *)

PRINT(200, X(C))
END.
```

Figure 4.7: A SIL program for solving a problem without an analytic solution.

The models in fig. 4.4 and 4.7 have much in common; their main difference being that two of the equations are different They both describe a dynamic DAE problem, and they both have one (linear) differential equation and one (non-linear) algebraic/implicit equation. This way of modelling 'time-dependent' static problems is artificial but convenient. In fig. 4.8 we show a SIL model which also solves the above problem; here the 'time-dependency' is modelled as a real time dependency. We will not call this model a dynamic model; it is a quasi-stationary or a time-dependent static model.

```
$TITLE=Time-dependent static problem
BEGIN

VARIABLE     X(0.9);
TIME C(-1:1);

(*  The equation  x^3 e^x = c^2
    must be satisfied for varying  c  *)

1.0E-6 := X*X*X * EXP(X) - C*C;

(*  Print the solution  X  versus  c   *)

PRINT(200, X)
END.
```

Figure 4.8: A quasi-stationary SIL model for $x^3 = c^2$.

The solution to this problem is shown in fig. 4.9.

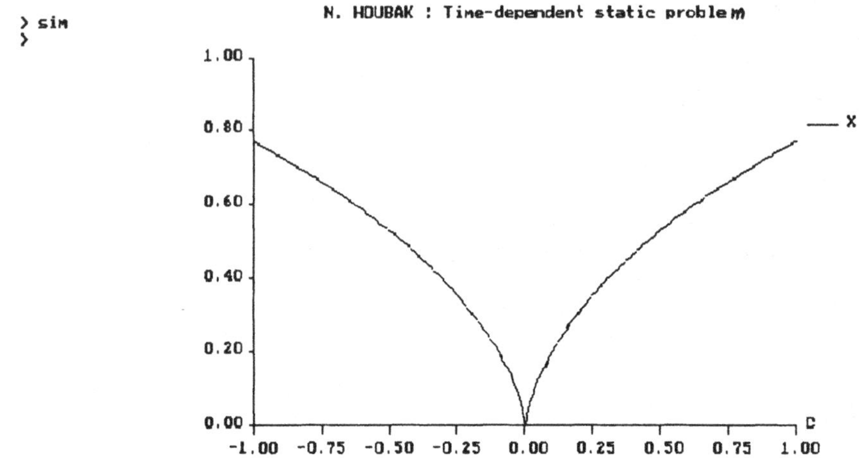

Figure 4.9: Solution to the quasi-stationary solution.

The next example illustrates the concept of several states in a model. The physical system is sketched in fig. 4.10, and the notation used refers to quantities defined in the figure.

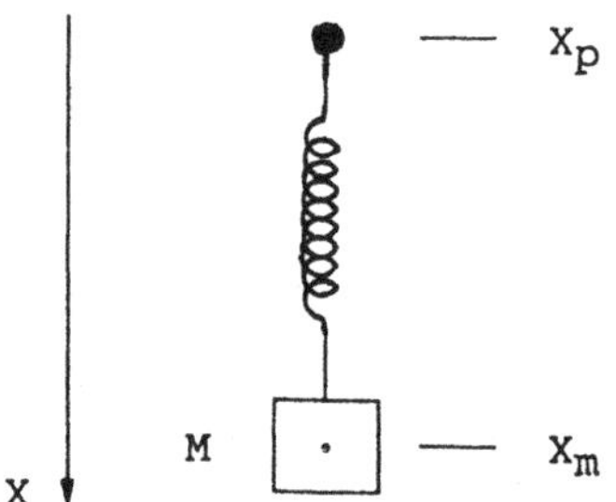

Figure 4.10: Physical model of a mass hanging in a spring.

The mass M is hanging in a spring of length L (unstretched). The position of M is called X_m, its derivative or the velocity is called V_m, and its derivative or the acceleration is called A_m. The spring-force F is defined as

$$F = k_1(L - l) \quad \text{for} \quad l > L$$
$$F = k_2(L - l) \quad \text{for} \quad l < L .$$

where l is the actual length of the spring, and k_1 and k_2 are two (different) spring constants. The spring is attached to a point with cyclic motion $X_p = A \sin(\omega t)$. This gives us the equation for l

$$l = X_m - X_p,$$

and Newton's second law applied to the mass gives us the equation for the acceleration

$$A_m = g + F/M ,$$

where g is the gravity constant. The initial velocity of M is 0 (zero) and the position $X_m(0)$ is computed such that the system is initially at equilibrium. The SIL model with appropriate values of the constants is shown in fig. 4.11.

```
BEGIN
PARAMETER  L(0.2),  K1(100),  K2(30),
           M(1.5),  G(9.82),
           A(0.08), OM(15);

VARIABLE   XM(),    XP,    LEN,    F;

DERIVATIVE VM(XM)(0),  AM(VM);

TIME T(0:10);

XM := L + M*G/K1;       (*  Initial value for  XM  *)

XP := A*SIN(OM*T);
LEN := XM - XP;         (*  Actual length of spring  *)

IF LEN > L THEN
   F := K1 * (L - LEN)     (*  Spring force  *)
ELSE
   F := K2 * (L - LEN);

AM := G + F / M;     (*  The differential equation  *)

PRINT(200, XP, XM, F)
END.
```

Figure 4.11: SIL model of a mass hanging in a spring.

Several examples of modelling discontinuous problems are given in chapter 6.2.4. For multi-state problems a diagram showing the different states and the possible transitions from one state to another may increase the insight in the problem. For each transition there normally is a condition on when the transition is performed; all the conditions must one way or the other appear in the SIL program as switching conditions. In fig. 4.12 we show the diagram for the previous example.

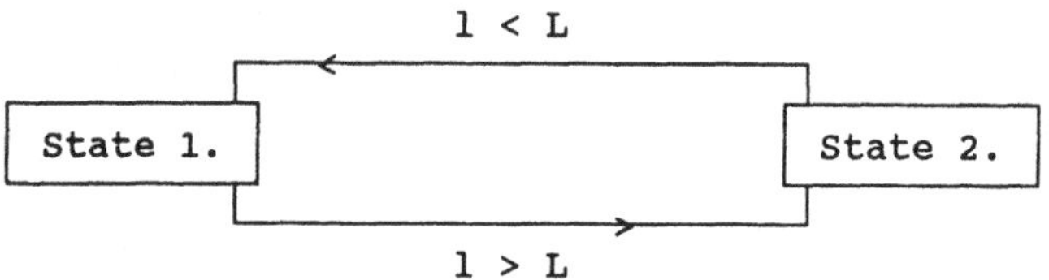

Figure 4.12: State diagram for the mass/spring problem.

For each state in the model there must be a corresponding block of statements in an IF THEN ELSE construction. The structure of the diagram reflects very clearly the structure of the nesting of the IF THEN ELSE statements in the SIL program.

In SIL it is a requirement that the model in all states defines the values of all variables. The reason for this is that no variable can be undefined on any part of the solution and the number of equations (both differential equations and algebraic equations) must remain the same throughout the whole solution process. This requirement is checked by the SIL compiler. The compiler also checks that the same variable is not used inconsistently in different states of the model; it is inconsistent when the same variable is used as an explicit variable in the THEN block and as an implicit variable in the corresponding ELSE block. The 'reverse' inconsistency is removed automatically by turning the explicit assignment into an implicit equation.

It is a very general characteristic of SIL models that there never is a statement like

```
X := X + 1 ,
```

the SIL language simply does not permit this kind of assignment. The very simple explanation for this is: the number of times that the model is evaluated is unpredictable before hand, therefore the value of X is unpredictable.

4.4. Model Limits in μSIL.

Though the μSIL system is very general in its concepts and can be used for solving problems in many different application areas, there are some limits that should be considered by the user. Those limits can be divided into two groups; the first kind of limitation is related to the types of problems that SIL is designed to solve, and the second kind of limitations is related to the implementation of the SIL system and the limits of the computer it is running on.

There are three types of problems for which SIL should not be used. These are partial differential equations (PDEs), differential algebraic equations (DAEs) with an index greater than or equal to 2 (two), and problems to be solved in real time.

PDEs normally give raise to a very large system of coupled ODEs when discretized and the current implementation of the μSIL system is not capable of handling large ODE systems in reasonable periods of time.

The SIL language will require the algebraic part of the DAE system to be structurally non-singular (no equations without implicit variables and no implicit variable without appearence in an algebraic equation). A singularity in value (index $\geq$ 2) is (normally) detected, and unless it only occurs at a single point it causes the simulation to stop.

Realtime problems cannot be handled correctly because the model is never evaluated in strictly monotone time; the ODE solver evaluates in each time step the model at selected points several times; also, a step may be reevaluated in case the error is too big.

The implementational limits on SIL models primarily reflect the size of the computer on which the system is running. The main memory is used both for storing a copy of the model in internal representation and for storing all the output points generated during the solution phase. On many DOS systems the main memory is limited to 640 Kbytes and this normally allows for some 200 Kbytes for the model and the output; for most models this will be quite sufficient.

There is a limit in the language on the number of output statements in a model; the μSIL system can handle at most 10 WRITE, PRINT, or PLOT statements, and none of them must specify more than 5 output items. The demo version of the system has some severe restrictions on the number of output points in each output statement. Also, the number of noise generators in a model is limited to 10. The number of sample time variables is limited to 10 whereas the number of discrete variables is only limited by the number of explicit static variables.

The μSIL system is always generated with a limitation on the maximum number of dynamic variables, implicit static variables, explicit variables, switch variables, and parameters because they reside in static workspace. Normally, these maxima are so high that the problem runs out of memory before any of the maxima is reached. The demo version of the system has a very low maximum (5) on most of these quantities.

The computing time in itself is not a limiting factor on the size of the models which can be run on the μSIL system but normally, one would prefer not to wait more than overnight to see the results of a simulation. Very often there is a relation between the size (and complexity) of a model and the time used for solving it but as demonstrated in chapter 3 the choice of solution method and also the choice of error tolerance have a major impact on the computing time used in the solution process. The speed of the computer also influences the speed of the solution process but buying a faster computer to solve a specific problem within a reasonable time should be considered the very last option.

The limits on the number of different types of variables illustrate what is considered a large problem in SIL: 50 dynamic variables of which up to 25 can have implicitly given derivatives, 25 implicit static variables, 250 explicit static variables, 50 switch variables, and 650 parameters. These numbers should not be interpreted in such a way that any model below these limits can be solved within a reasonably time on any PC. They reflect the fact that a simple model might be solveable even though it requires that the number of one of the above types of variables to be close to its maximum.

5. Numerical Methods.

The numerical methods used in the μSIL system will not be described in all details but most of the principles will be explained. It should be noted that the implemented methods are either standard software packages, or procedures written on bases of well known robust algorithms. There are in the present version of the μSIL system three basic numerical tasks involved; these are

1) solving linear systems of equations,
2) solving non-linear systems of equations, and
3) solving non-linear systems of ordinary differential equations.

The rest of this chapter contains a description of the solution algorithms used in the μSIL system.

The most basic numerical algorithm is the solution of linear systems of equations by some variation of Gaussian elimination. The problem is normally written as

$$\mathbf{A}\,\underline{x} = \underline{b}\ , \tag{5.1}$$

the matrix **A** contains the coefficients of the system, $\underline{x}$ is the vector of unknowns, $\underline{b}$ is the right hand side; the order of the matrix is N. The principle in Gaussian elimination is to split (factor) the matrix **A** into its **LU**-factors. The matrices **L** and **U** are computed by row- (or column-) operations, and they are lower and upper triangular matrices, respectively. Mathematically, the following equation holds,

$$\mathbf{A} = \mathbf{L}\cdot\mathbf{U}\ .$$

There are several variations of the factorization algorithm; the difference being mainly the order of the computation rather than a different formula. In the basic formula the element a_{ij} is updated at stage k by previously computed elements of the matrix,

$$a_{ij}^{(k+1)} = a_{ij}^{(k)} - a_{ik}^{(k)} / a_{kk}^{(k)} * a_{kj}^{(k)} \qquad k \le i,j \le N$$

The elements of the L matrix are the multipliers

$$l_{ik} = a_{ik}^{(k)} / a_{kk}^{(k)}$$

whereas the elements of the **U** matrix are

$$u_{kj} = a_{kj}^{(k)} \quad .$$

The numerical stability of the computations is ensured by requiring all elements of **L** being less than 1 ($l_{ik} \leq 1$). In case one of the diagonal elements of **U** becomes (in some sense) very close to zero, the matrix is considered (numerically) singular. This indicates that one of the equations is a linear combination of one or more of the other equations.

Having generated the **L** and **U** matrices it is easy to solve the system of linear equations (5.1) by direct computation;

$$\mathbf{L}\,\underline{z} = \underline{b}$$
$$\mathbf{U}\,\underline{x} = \underline{z} \; .$$

Since **L** and **U** are triangular matrices these two equations can be solved directly by (forward- and backward-) substitutions.

All mathematical software libraries have subprograms for solving systems of linear equations with different options. The matrix **A** can be either a real symmetric matrix, a real symmetric banded positive definite matrix (real positive eigenvalues), a tri-diagonal matrix, a sparse matrix (a large matrix with only few non-zero elements), a complex matrix, or a general real matrix. In μSIL all matrices are treated as general real matrices and they are stored in full storage mode. The interested reader can obtain further information about the solution of linear equations in Jennings (1985).

Solving systems of linear equations is a corner-stone in the task of solving a system of non-linear equations. This task is much more complicated for the following four reasons:

1) an iterative method (for example Newton's method) has to be applied,
2) a good initial approximation to the solution must be supplied,
3) there is no guarantee that the method will converge at all, and
4) there can be several (mathematical and numerical) solutions to the problem and the solution found need not be the one that is of interest.

In μSIL we use Quasi-Newton methods for solving the systems of non-linear equations involved. Solving a system of non-linear equations is mathematically formulated as finding a zero of a function, or

$$\underline{F}(\underline{x}) = \underline{0} \,. \tag{5.2}$$

The Newton formula for the n'th iteration then becomes

$$J(\underline{x}_n)\ \underline{\Delta x}_n = -\underline{F}(\underline{x}_n)$$
$$\underline{x}_{n+1} = \underline{x}_n + \underline{\Delta x}_n$$

where

$$J(\underline{x}_n) = \left\{ \frac{d}{d\underline{x}}\ \underline{F}(\underline{x}_n) \right\}$$

is called the Jacobian matrix. In each iteration this matrix must be updated and factorized in order to compute $\underline{\Delta x}_n$.

It is practice in Quasi-Newton methods to substitute the Jacobian matrix $J(\underline{x}_n)$ with an approximate matrix J. This could for example be the matrix used in the most previous iteration; thereby we avoid both the update and the factorization in each iteration. In case though the iteration converges too slowly the matrix must be updated and factorized.

It is important that the initial approximation $\underline{x}_0$ is as close as possible to the solution. The better the initial approximation, the fewer iterations are needed. Also, the convergence to one particular solution is only guaranteed when $\underline{x}_0$ is within a certain distance from that solution. It is worth mentioning that it can be proved, that if a Quasi-Newton method converges, it will converge to a solution of the original problem (5.2).

In case there are several (mathematical) solutions to a problem, the solution found depends on the initial guess; it should therefore be close to the solution of interest. For a model of a physical system it is normally the case that the 'physical' solution is the only solution within the range of the variables for which the model is valid. The μSIL system has built-in facilities for checking the validity of a solution and we strongly recommend that they be used.

There are two different criteria for stopping the iteration. The first one is to stop when the residual ($\underline{F}(\underline{x})$) is small; the second one is to stop when the displacement ($\Delta\underline{x}$) is small. Since the function $\underline{F}(\underline{x})$ can be arbitrarily scaled, the first criteria is not good alone. The second criteria also has its deficiencies; therefore, in µSIL the stopping criteria is that both the residual and the displacement must be small. A simulation is stopped in case the nonlinear equation system solver has failed to obtain a solution within a limited number of iterations (by default 250).

For some problems the Jacobian matrix $\mathbf{J}(\underline{x})$ can be a singular matrix. The SIL language normally prevents a structural singularity (a zero row or column in $\mathbf{J}(\underline{x})$). A singularity by value is normally caught by the linear equation solver. Both situations may indicate that an equation is missing or should be reformulated. The situation also arises when in the modelling phase two equations are used to generate a third equation; though it may look quite different from the other two it does not express anything new.

5.1. Solving Systems of Ordinary Differential Equations.

The most important part of the µSIL system is its ability to solve systems of ordinary differential equations (ODE's). This is a major task in almost all simulations. Below we give the requirements for a good ODE-solver.

1) It must be **reliable**; the results produced must be close to the 'true' mathematical solution to the problem. The philosophy is that wrong results are worse than no results even though they are computed at high speed.
2) It must must be **robust**; if given an ill-posed problem it is better to stop and indicate that something is wrong with the model instead of continuing the computations.
3) It must be **efficient**; normally, solving the system of ODE's is the most (CPU-) time consuming part of the simulation and it is therefore important that it is done as efficient as possible.
4) It must be **flexible**; it should be possible to interact with the ODE solver during the solution process and the user should have the possibillity of changing various parameters used for controlling the solution process.

5) It must be **user-friendly**; the μSIL system is an attempt to make a user-friendly interface to an integration routine.

The first 4 requirements have lead us to choose STRIDE made by Burrage, Butcher, and Chipman (1980) as the primary integration routine. On the other hand it is possible to implement other integration routines in the μSIL system with very little effort. In this chapter we will not describe STRIDE in itself, the interested reader is referred to its documentation, but we will go through some of the basic principles used in STRIDE as well as in other routines for solving stiff systems of ordinary differential equations.

In almost any situation concerning ordinary differential equations one assumes that the system be given in standard form,

$$\underline{y}' = \underline{f}(t, \underline{y}) \ , \qquad \underline{y}(t_0) = \underline{y}_0 \ , \qquad t_0 \leq t \leq T \ .$$

As such, we also call it an initial value problem (IVP). From the given initial value the problem is then to follow the solution until the final time T. Sometimes the solution at the endpoint is of interest, sometimes the behaviour of the solution over the whole time interval is of interest. By the way, we call t the time but it need not be physical time; it is the independent variable, the variable with respect to which the differentiation in $\underline{y}'$ is performed.

The system of ODEs is solved in a step by step manner; that is, we discretize the problem. In fig. 5.1 we show what this may look like. The solution is only computed at discrete points, the step points $t_1, t_2, .., t_n, t_{n+1}, ..$ and the solutions $y_1, y_2, .., y_n, y_{n+1}, ..$ are only approximations to the true solutions $y(t_1), y(t_2), ..,$ $y(t_n), y(t_{n+1}), ..$

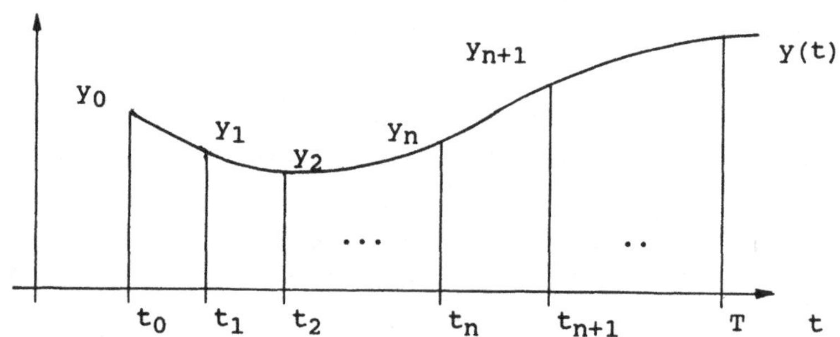

Figure 5.1: A discretized solution.

Advancing the solution from one point to the next is done by approximating the solution with either a polynomial or a rational function and thereby introducing some (hopefully minor) errors. Many will intuitively assume the derivative to be constant over a step (approximating the solution with a straight line), and assuming the stepsize to be small enough one could hope the solution to be correct. This method is called the explicit Euler method

$$\underline{y}_{n+1} = \underline{y}_n + h\,\underline{f}(t_n, \underline{y}_n) \ ,$$

where h is the (constant) stepsize. There are several good reasons for avoiding that method, they are

1) the method is explicit and of order one; on many problems this means that the stepsize should be very small if a solution has to be found with reasonably accuracy, or if the equations are stiff,
2) for small stepsizes the round-off errror committed in the calculation of the function $\underline{f}$ has a tendency of accumulating 'one-sided' in the solution; this has the effect that some problems can never be solved,
3) constant stepsize is normally considered a waste of CPU-time; a variable stepsize strategy makes the stepsize reflect the behaviour of the solution (long steps on a smooth solution and small steps on a rapidly varying solution), and they ensure the error of the solution to be within prescribed bounds,
4) the way explicit Euler is usually implemented together with a certain model makes it very hard to distinguish between the mathematical model and the numerical model; this makes it very difficult to determine whether a particular behaviour of the solution originates from the equations or from the numerical method.

A much better method with respect to numerical stability is the implicit (backward) Euler scheme, it goes like this

$$\underline{y}_{n+1} = \underline{y}_n + h\,\underline{f}(t_{n+1}, \underline{y}_{n+1}) \ .$$

The only (but very important) difference compared to the explicit Euler method is that the solution at the end-point of the step appears in the function f(t, y). This makes the method implicit and computationally more expensive. At each step we have to solve a system of non-linear equations of the form

$$\underline{F}(\underline{y}_{n+1}) = \underline{y}_{n+1} - \underline{y}_n - h\,\underline{f}(t_{n+1}, \underline{y}_{n+1}) = \underline{0} \tag{5.3}$$

by the Quasi-Newton method mentioned earlier. For many engineering problems though, this technique payes off. Notice, when a step is completed using the implicit Euler method the solution can always be extrapolated into the next step using the explicit Euler method (at no cost) because the derivative is already computed. This can be used to generate a good initial guess for the Newton iteration in the following implicit Euler step and the difference between the two solutions can be used in the error estimate (see later).

The stability properties of the two Euler methods can be illustrated by applying them to the so called test equation:

$$y' = \lambda \cdot y\ , \quad y(t_0) = y_0\ , \quad t_0 \le t \le T\ ,$$

where λ is a given constant. The exact solution to this equation is

$$y(t) = y_0 \cdot e^{\lambda \cdot (t-t_0)} \quad .$$

Applying the Euler methods to this equation gives

$$\text{E:} \quad y_{n+1} = y_n + h \cdot \lambda \cdot y_n \quad = y_n \cdot (1 + h \cdot \lambda) = y_0 \cdot (1 + h \cdot \lambda)^{n+1}$$

$$\text{I:} \quad y_{n+1} = y_n + h \cdot \lambda \cdot y_{n+1} = y_n / (1 - h \cdot \lambda) = y_0 \cdot \left(\frac{1}{1 - h \cdot \lambda}\right)^{n+1}$$

If λ is negative (the system is damped) we have from the analytical solution that $|y(t)| \le |y_0|$. The stability region (in the $h\lambda$-plane) for a numerical method is defined such that the same relation holds for all generated solutions; $|y_{n+1}| \le |y_0|$. For systems of equations λ represents the eigenvalues (the time-constants) of the system; therefore, λ is assumed to be complex. Fig. 5.2 shows the stability regions for both methods. Explicit Euler is stable <u>inside</u> the left circle and Implicit Euler is stable <u>outside</u> the right circle.

The figure indicates that problems with widely distributed eigenvalues (with negative real parts) require explicit Euler to use a very small stepsize in order to have all the $h \cdot \lambda$ values within its stability region. On the other hand, the implicit Euler method is stable for any stepsize used.

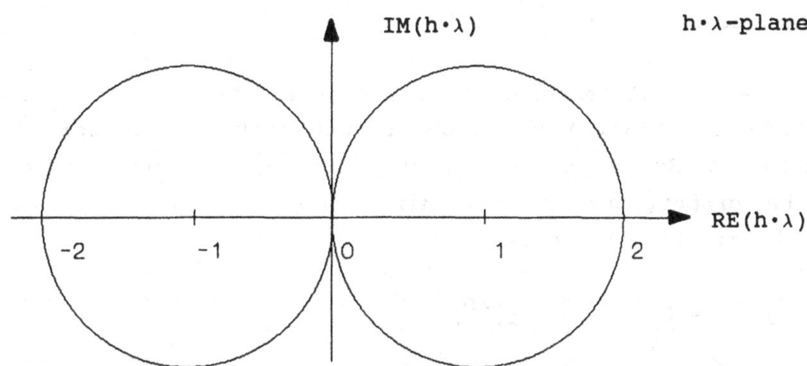

Figure 5.2: The stability regions for the explicit Euler method and for the implicit Euler method, respectively.

Some models give rise to higher order differential equations; for example, Newton's second law normally gives rise to second order equations. For a p'th order equation we have

$$y^{(p)} = f(t, y, y', .., y^{(p-1)}) .$$

The initial conditions necessary to give a unique solution are

$$y(t_0) = y_0, \quad y'(t_0) = y_0', \quad .., \quad y^{(p-1)}(t_0) = y_0^{(p-1)} .$$

The time is t and the time interval is $[t_0, T]$. By introducing extra variables for all lower order derivatives we can rewrite the problem as a system of coupled first order differential equations.

$$\begin{aligned} y_1' &= y_2 , & y_1(t_0) &= y_0 \\ y_2' &= y_3 , & y_2(t_0) &= y_0' \\ &.. & & \\ y_p' &= f(t, y_1, y_2, .., y_{p-1}) , & y_p(t_0) &= y_0^{(p-1)} . \end{aligned}$$

µSIL automatically transforms higher order differential equations to a system of first order equations using this technique.

When solving stiff systems of ordinary differential equations it is very important that the stepsize can vary, depending on the nature of the solution; small steps when the solution is rapidly varying and long steps when the solution is smooth. This is normally accomplished by estimating the error commited in each step by the numerical method and then adjusting the stepsize in order to make this error less than or equal to a user given bound.

There are several techniques for estimating the error of the method; see Lambert (1973). Here we shall just mention that for the explicit/implicit Euler scheme described above, the error after a step will be proportional to the difference between the two solutions (this is called Milne's device).

$$\underline{Err}_{n+1} = c_m \, (\underline{y}^{imp}_{n+1} - \underline{y}^{exp}_{n+1})$$

The constant c_m is equal to 0.5.

The stepsize h_{n+1} to be used in the next step can be computed on basis of the estimated error by the formula

$$h_{n+1} = h_n \cdot \sqrt[p+1]{\|\underline{Err}_{n+1}\| \; / \; EPS}$$

where p is the order of the method (the Euler methods are both of order 1), EPS is the user given error tolerance, $\|\underline{Err}_{n+1}\|$ is the norm of the error vector, and h_n is the current stepsize.

Since the steppoints due to the above stepsize strategy are unevenly distributed along the time axes, how do we then compute the solution at specific points, for example for output? The answer is: the solution is interpolated using information already computed. When the implicit Euler step is completed, the solution y_α at time $t_\alpha = t_n + \alpha \cdot h$, $\alpha \in [0, 1]$, can be computed by the formula

$$\underline{y}_\alpha = \underline{y}_n + \alpha \cdot h \cdot \underline{f}(t_{n+1}, \underline{y}_{n+1}) \; .$$

For higher order methods (for example like STRIDE) the interpolation formula is somewhat more complicated, but the solution can be interpolated to within the same order of accuracy as obtained by the numerical solution process. With an interpolation facility incorporated there is no need for the numerical method to actually hit specific points (in time); the stepsize strategy can therefore fully control the stepsize.

Many simulation problems are not 'simple' ordinary differential equations; they have an associated system of algebraic constraints which must be satisfyed all along the solution. This is what we call a differential algebraic equation (DAE) system. In general, these problems can be written as follows

$$\underline{0} = \underline{f}(t, \underline{y}, \underline{y}') , \quad \underline{y}(t_0) = \underline{y}_0 , \quad t \in [t_0, T] . \tag{5.4}$$

Very often though, it is possible to rewrite the problem into what is called semi-explicit form

$$\begin{aligned} \underline{0} &= \underline{g}(t, \underline{y}, \underline{z}) \\ \underline{y}' &= \underline{f}(t, \underline{y}, \underline{z}) , \quad \underline{y}(t_0) = \underline{y}_0 , \quad t \in [t_0, T] . \end{aligned}$$

The vector $\underline{z}$ contains the static implicit variables; that is, the variables whos values are determined on basis of the $\underline{g}$ function. Given in this form, the problem can be solved by the implicit Euler method simply by incorporating the $\underline{g}$ function into equation (5.3) and then using the Newton method to solve for both $\underline{y}_{n+1}$ and $\underline{z}_{n+1}$ in the same iteration process. The technique used in the μSIL system is that each time the derivative $\underline{y}'$ is computed, a Newton iteration first solves the $\underline{g}$-function with respect to $\underline{z}$ (t and $\underline{y}$ are given); the value of $\underline{z}$ is then used when computing the $\underline{f}$-function.

The system of algebraic equations (the $\underline{g}$-function) may not have a solution, it may have only 1 solution, or there may be several solutions. Even in the case where there is only 1 solution it may be difficult to find it; specially if the Jacobian matrix

$$J_g = \frac{\partial}{\partial \underline{z}} \underline{g}(t, \underline{y}, \underline{z})$$

is singular (at that solution point). We say that the index of the problem is greater than or equal to 2.

We can reduce the index of a problem by differentiating the equations (the $\underline{g}$-function) with respect to time and substitute the derivative of $\underline{y}$ ($\underline{y}'$) into the equations. Since all the algebraic equations must be equal to 0 at all times, their time derivatives must also be equal to 0 at all times. It is assumed that the J_g matrix is square. The SIL language also supports the definition and solution of problems with implicitly given derivatives, see equation (5.4).

Models defined in the SIL language can be discontinuous; that is, the SIL language allows the equations of the model to change at certain points in the solution process. If such models were solved uncritically with the above variable stepsize method very many steps will be wasted when passing the discontinuity. The way a discontinuity is defined in the SIL language makes it possible to always define a discontinuity function (a g-stop function) which passes through zero at the point (in time) where the discontinuity is located.

We can define a discontinuous problem as

$$\left.\begin{array}{lll} \underline{y}' = \underline{f}_1(t, \underline{y}) & \text{for} & g_1(t, \underline{y}) \geq 0 \\ \underline{y}' = \underline{f}_2(t, \underline{y}) & \text{for} & g_2(t, \underline{y}) \geq 0 \end{array}\right\} \quad \underline{y}(t_0) = \underline{y}_0$$

with the time $t \in [t_0, T]$. It is necessary to also give the initial state of the model in case it can not be determined uniquely on basis of the g-functions; in the above model both g_1 and g_2 can be greater than or equal to zero at the initial point.

After each step of the ODE-solver in μSIL all the 'active' g-functions are checked; if a change of sign is observed in one of the g-functions, μSIL activates a procedure for locating the root (in time) of that function. In case several g-functions have roots in the same step, only the root closest to the start of the step is considered. Evaluation of the discontinuity function (the g-function) is done in the following steps:

1) the solution is interpolated at the point,
2) the model (the $\underline{f}$-function) is evaluated, and
3) the g-function itself is evaluated.

This allows all variables of the model to appear in the g-functions.

The method used for finding the root of the g-function is basically a Regula-Falsi method. It exploits the fact that we have a point on each side of the root with opposite sign of the g-function; initially these are t_n and t_{n+1}, respectively. Since the g-function at those points is known, it can be approximated with a straight line. The intersection between the line and the axis becomes a new approximation $t_{r,i}$ to the root t_r. Based on the sign of the g-function at $t_{r,i}$, the interval around the root is reduced by discarding one of the previous end-points. The iteration is stopped when the interval around the root is small enough; special actions during the iteration ensure that both end-points are moved towards the root. The algorithm is visualized in fig. 5.3.

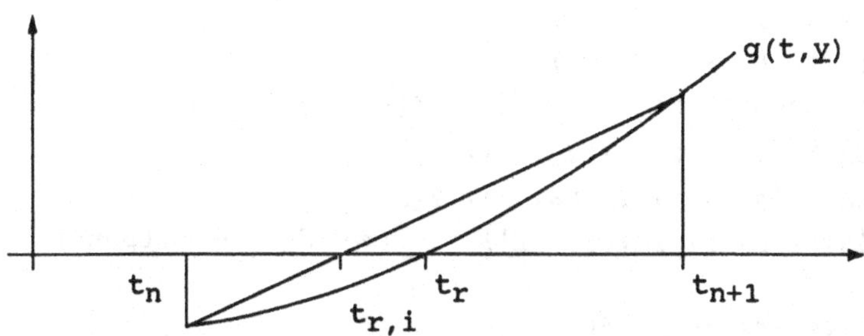

Figure 5.3: The Regula-Falsi root searching algorithm.

In order to describe the way a general model is evaluated in the μSIL system we define the functions $\underline{f}_1$, $\underline{f}_2$, .., $\underline{f}_9$. Each of these functions contains a part of the model equations and formally they all have identical structure with respect to discontinuities; the structure is dictated only by the structure of the SIL model and the only deviation is caused by the removal of 'empty' IF THEN ELSE statements.

With the notation

- $\underline{p}$ for the parameters,
- $\underline{y}_e$ for static explicit variables,
- $\underline{y}_i$ for the static implicit variables,
- $\underline{r}_i$ for the residual of implicit equations,
- $\underline{y}$ for the dynamic variables,
- $\underline{y}'$ for the explicit derivatives,
- $\underline{y}_i'$ for the implicit derivatives, and
- $\underline{r}_d$ for the residual of implicit derivative equations,

the general setup for a model evaluation can be written in the following way:

Initially, the system evaluates the equations

$$\underline{y}_e = \underline{f}_1(\underline{p}, \underline{y}) ,$$

and when necessary during the integration the following dynamic part is evaluated,

$$
\begin{aligned}
&\underline{y}_e = \underline{f}_2(t, \underline{p}, \underline{y}, \underline{y}_e)\\
&\quad \text{At each Newton iteration} \quad (\underline{f}_4(t, \underline{y}_i) = \underline{0})\\
&\quad \underline{y}_e = \underline{f}_3(t, \underline{p}, \underline{y}, \underline{y}_e, \underline{y}_i)\\
&\quad \underline{r}_i = \underline{f}_4(t, \underline{p}, \underline{y}, \underline{y}_e, \underline{y}_i)\\
&\underline{y}_e = \underline{f}_5(t, \underline{p}, \underline{y}, \underline{y}_e, \underline{y}_i)\\
&\underline{y}' = \underline{f}_6(t, \underline{p}, \underline{y}, \underline{y}_e, \underline{y}_i)\\
&\quad \text{At each Newton iteration} \quad (\underline{f}_8(t, \underline{y}') = \underline{0})\\
&\quad \underline{y}_e = \underline{f}_7(t, \underline{p}, \underline{y}, \underline{y}_e, \underline{y}_i, \underline{y}_i')\\
&\quad \underline{r}_d = \underline{f}_8(t, \underline{p}, \underline{y}, \underline{y}_e, \underline{y}_i, \underline{y}_i')\\
&\underline{y}_e = \underline{f}_9(t, \underline{p}, \underline{y}, \underline{y}_e, \underline{y}_i, \underline{y}_i') \quad \text{(mainly for output.)}
\end{aligned}
$$

This structure ensures that

1) those variables that do not vary during a simulation are only evaluated once,
2) some (specially defined) explicit variables can be used as auxillary variables during the Newton iteration,
3) output variables can depend ($\underline{f}_9$) on all other variables,
4) implicit algebraic equations and implicit differential equations can be mixed in the same model, and
5) explicit and implicit differential equations can be mixed in the same model.

In case a model defines, for example, a derivative or an implicit variable to be an output variable, the model is evaluated at each output point. If the output statements only contain dynamic variables no extra evaluations of the model will be performed.

This was a description of algorithms currently implemented in the μSIL system. In future versions of the system it is planned to have algorithms for

1) interpolation of user tables,
2) optimization,
3) fast Fourier transforms of output data tables,
4) numerical integration of finite integrals, and
5) computing eigenvalues (and eigenvectors) of the Jacobian matrices.

6. The SIL Language.

The SIL language is the corner stone in the whole μSIL system because all model information must be specified using this language. One of the ideas behind this language is that you have programming languages for making programs, so why not have a simulation language for making simulation models. This also has the advantage that many well known techniques from building compilers for programming languages can be used when implementing the simulation language compiler.

Common to most computer languages is a syntax. Most languages have a very strict syntax, meaning that the compiler requires very specific rules to be followed in order to be able to handle the user program; the SIL language compiler is no exception. Most of this chapter is devoted to the description of the syntax of the SIL language. We use the very formal but precise Backus-Naur notation (BNF) for this; hopefully this should not cause too many troubles for the reader and the examples with explanations included normally cover the most important parts.

The semantics of the various parts of the language is explained simultaneously with the syntax and the final two sections explain more about the semantics and how to build models in the language.

The language is designed to be a whole-in-one language; that is, no interface has been designed that allows any part of the model to be defined outside the language. This design has the advantage that the μSIL system at any time has full control over the model. This is essential for an efficient treatment of discontinuities. It also makes it easier to detect and handle errors in the model. This philosophy also requires that the language has many facilities and options for different tasks; to fulfil this, the language must be easily extendable and future versions of the program will also contain additions to the language. The basis for the language is as defined in this chapter and the extensions will never influence how this basis works; they can only increase its possibilities.

In programming languages the type of a variable very much reflects the underlaying representation in the computer. We distinguish between INTEGERs and REALs because their internal representation and hence the operations performed on them are very different. In the simulation language SIL the type of a variable reflects what the variable can be used for; all variables are stored internally as "REALs" (except some control variables). Another major difference between SIL and a programming language is that a 'normal' program is executed sequentially from the top; loops in the language can be used to force a repeated execution of selected statements. In SIL, the program is a model description and is evaluated (from top to buttom) each time the model solver may require it; the SIL compiler may rearrange the order of computation of the statements but the SIL language in itself is loopfree.

The SIL language was initially intended as a tool for structuring the process of building minor simulation models, and coupled with the other facilities of the μSIL system it is an easy way to obtain convinient solutions of models, too. It is a common experience that the coupling between a mathematical model and a graphical representation of the solution is very important for the understanding of models. There are several bottlenecks in the μSIL system which prevent very large models from being solved efficiently. On the other hand, the methodology learned by using the SIL language makes it much easier to use or to design software for solving large and complex simulation problems. Minor models are also often better understood even if they are more complex, and therefore they give better insight in the behaviour of the physical system.

6.1. Fundamentals.

Before we go into details in the SIL language we shall look at some basic elements in almost every "computer language". We will extensively use the BNF-form (Backus Naur form) to describe the complete syntax of the SIL language. We will give a complete explanation of both the syntax and the semantics of the language and give some examples of how the different elements can be used.

Every compiler normally reads the written source text character by character. The SIL character set consists of the digits '0' to '9' (<digit>), all letters, capital and small, from 'a' to 'Z' and underline '_' (<letter>). Capital and small letters are not distinguished by the compiler. Beside these characters, special characters like period, comma, colon etc. are also recognized.

We can now define what we will understand by an identifier (<id>) of (say) a variable

$$\langle\text{id}\rangle ::= \langle\text{letter}\rangle \left\{ \begin{matrix} \langle\text{letter}\rangle \\ \langle\text{digit}\rangle \end{matrix} \right\}_0^*$$

An identifier starts with a letter followed by an arbitrary number of letters and digits without blanks or any other special character. In SIL two identifiers must differ within the first 11 characters to be considered different. An identifier can not be divided between one line and the next.

Another important building block is a number (<number>), which is defined as :

$$\langle\text{number}\rangle ::= \left\{ \begin{matrix} \langle\text{real number}\rangle \left\{ \text{'} \left\{ \begin{matrix} + \\ - \end{matrix} \right\}_0^1 \{\langle\text{int}\rangle\}_0^1 \right\}_0^1 \\ \text{'} \left\{ \begin{matrix} + \\ - \end{matrix} \right\}_0^1 \{\langle\text{int}\rangle\}_0^1 \end{matrix} \right\}$$

$$\langle\text{real number}\rangle ::= \left\{ \begin{matrix} \langle\text{int}\rangle \{ .\ \{\langle\text{int}\rangle\}_0^1 \}_0^1 \\ \{\langle\text{int}\rangle\}_0^1\ .\ \{\langle\text{int}\rangle\}_0^1 \end{matrix} \right\}$$

$$\langle\text{int}\rangle ::= \langle\text{digit}\rangle \{\langle\text{digit}\rangle\}_0^*$$

Note that ' is used as exponent mark, but E and D may be used as well. Unfortunately this gives some complications. For instance is E12 an identifier (<id>), whereas '12 is a real number, and .D3 and '3 are the same real number. It is strongly recommended to always use ' especially because the internal representation does not depend on the exponent symbol!

Arithmetic expressions are built according to the following syntax which ensures the normal hierarchy of operators.

$$\text{<expression>} ::= \left\{ \begin{matrix} + \\ - \end{matrix} \right\}_0^1 \text{<term>} \left\{ \left\{ \begin{matrix} + \\ - \end{matrix} \right\} \text{<term>} \right\}_0^*$$

$$\text{<term>} ::= \text{<factor>} \left\{ \left\{ \begin{matrix} * \\ / \end{matrix} \right\} \text{<factor>} \right\}_0^*$$

$$\text{<factor>} ::= \text{<primary>} \{ \text{** <int>} \}_0^1$$

$$\text{<primary>} ::= \left\{ \begin{matrix} \text{<noise generator>} \\ \text{<number>} \\ \text{<id>} \\ \text{<std function designator>} \\ \text{(<expression>)} \end{matrix} \right\}$$

$$\text{<noise generator>} ::= \text{NOISE} \{ (\{\text{<seed>}\}_0^1 \{, \text{<type>}\}_0^1) \}_0^1$$

$$\text{<seed>} ::= \text{<int>}$$

$$\text{<type>} ::= \text{<int>}$$

$$\text{<std function designator>} ::= \text{<std function id>} \text{ (<expression>)}$$

$$\text{<std function id>} ::= \left\{ \begin{matrix} \text{SIN} \\ \text{COS} \\ \text{ATAN} \\ \text{EXP} \\ \text{LOG} \\ \text{SQRT} \end{matrix} \right\}$$

The operator with highest priority is the power operator. The current syntax states that only positive integer powers can be used. If for some reason real powers are needed the user must rewrite that part of the expression using the EXPonential and LOGarithm functions. In future versions of the μSIL system this option may be included.

The noise generator is essentially a random number generator. It should be used with much care because it introduces some uncertainty into the model. A noisy variable should not be used for the determination of a state shift (a discontinuity). <seed> is an integer giving the seed for the generator; if omitted, an artificial seed will be generated.

<type> indicates the type of the noise generator:

= 1 gives a uniform distribution of numbers between 0 and 1 scaled with 'current step'/MAXSTEPSIZE,

= 2 gives a normal distribution (mean value = 0, and variance = 1) of numbers with the above scaling,

= 3 as for <type> = 1, but in [-1, 1].

= 4 gives the uniform distribution [0, 1] without scaling,

= 5 gives the normal distribution without scaling, and

= 6 gives the uniform [-1, 1] distribution without scaling.

The elements <number>, <id>, and <std function designator> are the possible altenatives for an operand in an <expression>, whereas (<expression>) allows a sub-expression to be put into parenthesis; this is necessary when the order of evaluation is important (for example (A+B)*C).

It is possible to put comments in the source text. They can be written everywhere where a blank is legal. The comment must be surrounded by (* and *) in that order .

An important concept is the reserved words. They are built as identifiers (<id>) but the compiler knows them and treats them in a special way. Reserved words are not allowed as names of variables etc. Below we give a complete list over the reserved words:

BEGIN	END				
VARIABLE	DERIVATIVE	PARAMETER	MACRO	SWITCH	
TIME	SAMPLETIME	DISCRETE			
IF THEN	ELSE	AND	OR	ON	OFF
FOR	STEP	UNTIL	DO		
WRITE	PRINT	PLOT			
SIN COS	ATAN	EXP	LOG	SQRT	NOISE
ABSERROR	RELERROR	STEPSIZE	MAXSTEPSIZE		
METHOD	MAXORDER	MAXCPU	INTEGRATOR		LANGUAGE

The reserved words as well as identifiers must be separated from the rest of the source text with at least one special character (e.g. a space), which is not used in an identifier. For example A12.34 will be detected as the identifier A12 followed by the number 0.34.

6.2. Statements.

When one simulates a system one has to describe the corresponding model in an appropriate language. SIL is ALGOL- or PASCAL- like, whereas other simulation packages use more or less specialized languages. In SIL we describe the model with a program which by definition is a <statement> followed by a period '.'. Formally in BNF :

```
<program> ::= <statement> .
```

A statement plays a central role in all block oriented languages. Blocks are defined as follows in recursive BNF :

```
<block> ::= BEGIN <statement list> END
                                                   *
<statement list> ::= <statement> { ; <statement>}
                                                   0
<statement> ::= <block>
```

Typically that gives the following structure of a program :

```
BEGIN
{ statements }
END.
```

Besides giving structure in the language, a statement is used to define variables (<id>) and macros, to express the equations of the model or to activate the expansion of a macro. We can write :

```
                  ( <block>                )
                  | <declaration>          |
                  | <assignment>           |
<statement> ::=  {  <switch assignment>     }
                  | <if then else>         |
                  | <macro call>           |
                  ( <output statement>     )
```

In the SIL language all variables **must** be declared before they are used. Declaration statements can be placed anywhere in a SIL program but it is good practice to declare all variables in the very beginning of the program. The simplest and the most important "model statement" is the assignment. The equations of the model are normally formulated with this statement.

Many models operate in several states; using IF THEN ELSE statements is a very structured way of defining such problems. This requires the user to be very careful when making such models; which conditions are the most appropriate and how should they be modelled in SIL. Chapter 6.2.4 addresses this problem. Closely related to the IF THEN ELSE construction are the SWITCH (or 'LOGICAL') variables and their assignment. The variables themselves normally take the values ON or OFF but they always represent a relation between two expressions (or variables). Internally this relation is represented by a function (a discontinuity function) which changes sign whenever the value of the SWITCH variable changes.

In SIL, a macro can be used to make a model of a general component (a pump, a transistor, etc.). The macro is called (expanded) whenever the component is needed. The parameter transfer to the macro is used for specifying the values of the parameters. Macros are specially useful in cases where the same sub-model should be applied several times (maybe with slightly different values of some parameters).

Output statements are used for specifying which variables should have their values 'sampled' during a simulation (collected at time-equidistant points) and displayed. Also, the type of display is determined by the output statements. The output processing is so important that SIL will not accept a model without an output statement.

6.2.1. Declarations.

The purpose of the declaration of variables is to specify the exact meaning of an identifier for the rest of the program. The SIL compiler permits the mixing of declarations with the other types of statements, but it is practical to put the declarations all together at the beginning of the program. In order to make it possible for the compiler to recognize declarations they must start with one of the following seven reserved words:

PARAMETER	TIME	VARIABLE	DERIVATIVE	SWITCH
MACRO	DISCRETE	SAMPLETIME		

These keywords tell the SIL compiler how to treat the rest of the statement. We can split the description of declarations accordingly :

$$\text{<declaration>} ::= \left\{ \begin{array}{l} \text{<parameter declaration>} \\ \text{<time declaration>} \\ \text{<variable declaration>} \\ \text{<derivative declaration>} \\ \text{<switch declaration>} \\ \text{<macro declaration>} \\ \text{<discrete declaration>} \\ \text{<sampletime declaration>} \end{array} \right\}$$

and we will describe the declarations of the different types of variables one at a time.

6.2.1.1. PARAMETER.

If a given model contains a quantity which either is assumed to be constant or by definition is constant, it may (and should) be declared as a parameter. Typically, quantities like reference pressure, density, gravity, material parameters etc. can be declared as parameters. The syntax for the parameter declaration is:

$$\text{<parameter declaration>} ::= \text{PARAMETER <par id>} \{, \text{<par id>}\}_0^*$$

$$\text{<par id>} ::= \text{<id>} \text{ (<initval>)}$$

$$\text{<initval>} ::= \left\{ \begin{matrix} + \\ - \end{matrix} \right\}_0^1 \text{<number>}$$

A program with nothing but parameters is shown here:

```
BEGIN
PARAMETER CP(1.6'-3),  FLOW(1E-3);
PARAMETER TMAX(22        (* degrees C *) ),
          TMIN(17.5      (* degrees C *) );
PARAMETER LENGTH(10)     (* in CM *),
          HEIGHT(20)     (* in CM *);
PARAMETER ATM_PRES(1.01325)  (* in Bar *);
END.
```

As mentioned earlier everything between (* and *) is a comment; in this example it is used to give the units of the different quantities. As one can see in the syntax, the parenthesis with the value of the parameter must always follow the name of the parameter, blanks in between are permitted.

A characteristic of a parameter is that when simulations are performed from the graphic command system one may change the value of any identifier defined as a parameter without having to recompile the whole model; this makes it easy to see the influence of the parameter on the solution.

6.2.1.2. TIME.

In some dynamic models there might be a need for accessing the "model time". For instance, the model may contain functions of time like solar influx, fuel consumption etc. With the TIME declaration one and only one time variable can be declared and it will always be the model time. The syntax is:

$$\langle\text{time declaration}\rangle ::= \text{TIME } \langle\text{id}\rangle \left\{ \left(\langle\text{range}\rangle \right) \right\}_0^1$$

$$\langle\text{range}\rangle ::= \left\{ \begin{array}{l} \left\{\begin{matrix}+\\-\end{matrix}\right\}_0^1 \langle\text{number}\rangle : \left\{ \left\{\begin{matrix}+\\-\end{matrix}\right\}_0^1 \langle\text{number}\rangle \right\}_0^1 \\ \qquad\qquad : \left\{\begin{matrix}+\\-\end{matrix}\right\}_0^1 \langle\text{number}\rangle \end{array} \right\}$$

Besides the declaration of the variable, an interval can be specified for the independent variable. If no interval is specified or there is no time declaration, the default time interval will be assumed. If the model to be solved is a dynamic model (it consists of at least one differential equation) the default interval is [0,1]; for static models the time is assumed to be 0 (zero) by default. It should be noted that the lower bound of the time interval must be smaller than the upper bound but the lower bound need not be 0 (zero). Here is an example of a TIME declaration.

```
TIME T(-1:22.5)
```

6.2.1.3. VARIABLE.

Essential components in a mathematical model of a system are its state variables or system variables. The declaration of such variables can be made as :

$$\langle\text{variable declaration}\rangle ::= \text{VARIABLE}\ \langle\text{var id}\rangle\ \{,\ \langle\text{var id}\rangle\}_0^*$$

$$\langle\text{var id}\rangle ::= \langle\text{id}\rangle \left\{ \left(\left\{ \begin{array}{c} \langle\text{range}\rangle \\ \langle\text{initval}\rangle \\ <\ \langle\text{tol}\rangle \end{array} \right\}_0^1 \right) \right\}_0^3$$

$$\langle\text{tol}\rangle ::= \langle\text{number}\rangle$$

We will in this section describe the different types of variables, how they are declared, what the different items of a declaration means and how they are used.

All variables may be assigned a range; that is, if during the solution of the model SIL encounters a variable that is outside its defined range, the solution is immediately terminated with an error message saying which variable that has had a failure in its range check. The default range (which is not checked for by SIL) for any variable is the whole range of numbers defined by the arithmetic of the computer.

Initial values must be supplied for dynamic variables, for implicit static variables, and for discrete variables. For the dynamic variables the initial value is the initial value for the integration; for the implicit static variables the initial value is used as the initial approximation in the Newton iteration; for discrete variables it is their first value. The initial value for a dynamic variable is directly changeable in the graphic command system. This makes it easier to perform several simulations with varying initial values without having to compile the whole model again.

For implicit static variables the initial value is used as the first start guess for the very first Newton iteration invoked to solve the implicit (non-linear) system of algebraic equations. For the following Newton iterations SIL will use the most recent solution as initial approximation. For some problems it could be necessary to limit the stepsize (MAXSTEPSIZE) in order to assure the convergence of the iteration. Changing the initial guess of the solution can result in SIL finding another solution to the model. Mathematically, both solutions are valid but the user must determine which one is the physical solution. We strongly recommend that implicit static variables and implicit derivatives always have a range associated with them.

Discrete variables also need an initial value since their values normally are determined by some kind of recursion scheme. Changing the initial value of such a variable can sometimes have dramatic effect on the series of values produced by the associated recursion.

The syntax allows an empty parenthesis to follow the declaration of a variable; this is used for two purposes. It can be used to declare a variable that is either a dynamic variable or a discrete variable and then indicate that the associated initial value is specified later in an assignment statement. This allows initial values to depend on some of the parameters in the model; this facility is described later in this chapter. The other purpose for using an empty parenthesis is to declare a special type of explicit static variable. The variable is supposed to depend on some of the implicit static variables or implicit derivatives and it will appear in some of the implicit equations; this requires the variable to be updated in each Newton iteration performed on the systems of implicit equations.

Dynamic variables can individually have an error tolerance associated with them. This is done by applying '< <number>' in the parenthesis. Specially, when the variables differ in magnitude it is difficult to adjust the predeclared variables ABSERROR and RELERROR in such a way that both a reasonably good overall accuracy is obtained and the computing time is minimized. The number given after the '<' sign will be the absolute error bound for that particular variable used in the integration routine; it must therefore be positive. See the chapter 6.2.1.4 DERIVATIVE for an example.

Variables declared without an initial value or identifier specification are treated as explicit static variables; they can eventually have a range specification. Explicit static variables can be used to store intermediate results but may also be a major part of a model. If an explicit static variable only depends on parameters, SIL automatically ensures that this variable is only evaluated once. The variable is treated almost as a parameter; the major difference is that in the graphic command system it is impossible to change the value of explicit static variables.

The simplest form for a variable declaration is the declaration of the static explicit variable, say Y.

```
VARIABLE  Y
```

The next appearence of Y must be on the left hand side of an assignment (:=); the statement looks like this

```
Y := ..........
```

and by evaluating the expression on the right hand side SIL can determine the value of Y. The expression on the right may contain either parameters, the model time, dynamic variables, or discrete variables. Also explicit static variables, explicit derivatives, implicit static variables or implicit derivatives are allowed, provided they already have been assigned a value. Since this is the first appearence of Y following its declaration, none of the variables on the right hand side depends on Y (directly or indirectly). Since Y for the rest of the model is assumed to maintain the value just computed, Y is no longer allowed to appear on the left hand side of an equation. The only exception to this rule is: If the assignment appears in an IF THEN ELSE construction, it must appear in both the THEN branch and the ELSE branch.

If we change the declaration to the following

```
VARIABLE Y(-3:6.5E2)
```

we may use the variable Y as described above with the exception that μSIL will interrupt the simulation if Y deviates from the interval [-3,650].

A declaration of the variable Z could look like this

```
VARIABLE Z (2.5)(0:)
```

Z can now be used for one of two purposes

1) Z can be a dynamic variable with initial value 2.5 and be non-negative. It will later appear in the declaration of a derivative variable.
2) Z can be a discrete variable with initial value 2.5 and it must always be non-negative. In this situation Z will later appear in a DISCRETE declaration.
3) Alternatively, Z can be an implicit static variable for which the value 2.5 should be used as initial approximation when solving the associated system of non-linear equations. Also in this situation Z will be checked by SIL for being non-negative.

If Z is either a dynamic variable or a discrete variable, we may as well declare it in the following manner:

```
VARIABLE Z()(0:)
```

After its associated derivative or 'next value' variable has been declared (see the next chapters), a statement like

```
Z := 2.5
```

may then be used in order to generate the appropriate initial condition. The right hand side of this statement could have been any valid arithmetic expression involving any parameter defined so far. This way of declaring the variable Z is also used for the special explicit static variables which take part in the solution of the system of non-linear equations. A small SIL program illustrating this facility is shown below.

```
BEGIN
VARIABLE   Y(1.2), Z( );
Z := EXP(Y);
1.0E-5 := Y + Z;
WRITE(Y,Z)
END.
```

The variable Z will be updated in each iteration of the Newton process when solving the equation

$$y + \exp(y) = 0 \quad .$$

The iteration is stopped when the residual is less than 10^{-5}.

6.2.1.4. DERIVATIVE

This variable type is used to establish a connection between a dynamic variable and its derivative with respect to the independent variable (usually the time). The syntax is :

$$\langle\text{derivative declaration}\rangle ::= \text{DERIVATIVE}\ \langle\text{der id}\rangle\ \{,\ \langle\text{der id}\rangle\}_0^{*}$$

$$\langle\text{der id}\rangle ::= \langle\text{id}\rangle\ (\ \langle\text{der of}\rangle\ )\ \left\{ (\ \left\{ \begin{matrix} \langle\text{initval}\rangle \\ \langle\text{range}\rangle \end{matrix} \right\}_0^1) \right\}_0^2$$

$$\langle\text{der of}\rangle ::= \langle\text{id}\rangle$$

Note that despite the syntax, the order of (<der of>), (<range>) and (<initval>) does not matter. If the variable in (<der of>) has the type DERIVATIVE and an initial value associated, then the declared derivative becomes the second derivative of a variable. This can be used for third or higher order derivatives as well. A little program illustrates this :

```
BEGIN
VARIABLE  Y(0)(<'-4);
DERIVATIVE DY (Y) (1),  D2Y(DY)(0:);
D2Y := -Y;
PRINT(Y(DY))
END.
```

Here the dynamic variable Y is declared, it is given the initial value 0 (zero), and a local error tolerance of 10^{-4} is associated. The derivative (y') DY of Y with initial value 1 is declared as well as the second derivative (y") D2Y with the condition that it should be non-negative. The declaration of D2Y makes the SIL compiler consider DY as a dynamic variable and no longer as a

DERIVATIVE variable. The following SIL program solves precisely the same problem but the rewriting to first order equations is done explicitly.

```
BEGIN
VARIABLE  Y(0)(<'-4), Y1(1);
DERIVATIVE DY (Y),  D2Y(Y1)(0:);
DY  := Y1;
D2Y := -Y;
PRINT(Y(Y1))
END.
```

Normally, derivatives appear explicitly in the equations but for some problems the derivatives cannot be given in closed form; they are given implicitly. In SIL this is handled in much the same way as for implicit static variables. When the derivative is declared an initial value is added; later, when the derivative appears in an implicit equation this becomes the implicit derivative equation. The special form of explicit static variables mentioned in section 6.2.1.3 can also be used in connection with implicit derivatives. This is illustrated in the following example.

```
BEGIN
VARIABLE Y(1), TEMP();
DERIVATIVE DY(Y)(2.6);
TIME T(0:3.5);

TEMP := DY*DY;
1.0E-5 := DY*EXP(TEMP) - TEMP;

WRITE(Y,DY,TEMP)
END.
```

A SIL model need not have any declared derivatives. If there are no derivatives in a model we call it a static model. Though a static model normally is time-invariant some kind of time-variation might be appropriate. Suppose a static model is given and it can be solved. The interesting thing about that model though, is how the static solution changes when one of the parameters is changed. One way of modelling this, is to introduce a differential equation

$$y' = 1$$

with an appropriate initial value such that its solution (y(t)) coinsides with the range of the parameter to be changed. All occurences of the varying parameter in the model are exchanged with y(t); the problem has become a differential algebraic problem. Alternatively, a TIME variable could have been used.

6.2.1.5. SWITCH.

This type of variable is used when modeling a system with several states. One can use these variables as models of switches, contacts or relays etc. Associated with each switch is a condition stating when the switch shifts from its present value to the other value. How this works will be explained later (chapter 6.2.4).

The syntax for declaring switch variables is :

$$\langle\text{switch declaration}\rangle ::= \text{SWITCH } \langle\text{switch id}\rangle \ \{,\ \langle\text{switch id}\rangle\}_0^*$$

$$\langle\text{switch id}\rangle ::= \langle\text{id}\rangle \ \{\ (\ \langle\text{switch initval}\rangle\ )\ \}_0^1$$

$$\langle\text{switch initval}\rangle ::= \left\{ \begin{matrix} \text{ON} \\ \text{OFF} \end{matrix} \right\}$$

An example of a program with declarations of switches is :

```
BEGIN
SWITCH RELAY_1, RELAY_2(OFF);
SWITCH STATE(ON);
  :
  :
END.
```

If an initial value is not given it must be calculable from the initial values of the variables in the model. See chapter 6.2.4 for the assignment of switches.

6.2.1.6. MACRO.

In a simulation model the same (or almost the same) component or set of equations may appear several times. Maybe some parameters change from time to time but essentially the same equations have multiple occurencies in the model. It is valid to enter the equations at each appearence but it is easier to define the equations as a MACRO and then let SIL expand the equations at each appearence. The MACRO is formally defined as follows:

```
<macro declaration> ::= MACRO <macro heading> ; <statement>

                                                     1
<macro heading> ::= <id> { ( <formal parm list> ) }
                                                     0

                                                        *
<formal parm list> ::=  <formal parm> {; <formal parm>}
                                                        0

                      ⎧ ⎧ VARIABLE   ⎫              *  ⎫
<formal parm> ::=     ⎨ ⎨ DERIVATIVE ⎬ <id> {, <id>}   ⎬
                      ⎪ ⎩ PARAMETER  ⎭              0  ⎪
                      ⎩ TIME <id>                      ⎭
```

Within a macro the only variables known are those declared in the macro and the variables passed via the parameter list. The statements of the macro are stored exactly as written in internal format and each time the macro is called, the statements replace the macro call. The actual variables in the macro call are substituted into the macro statements at the time of replacement.

Macros may be defined within macros. Notice, that a macro is only known within the macro in which it is declared; this ensures that a macro cannot be called recursively. The variables declared in a macro are only known within that macro; the use of global variables is not allowed.

6.2.1.7. DISCRETE and SAMPLETIME.

In order to be able to define discrete event systems we need three different types of variables. First, the discrete solution itself is stored in a variable of type VARIABLE, and it must have an initial value associated. Second, the sample time interval must be specified in a variable of type SAMPLETIME; it can be either constant or varying. The third type of variable needed is DISCRETE variables; these variables are used for computing the next value of the associated discrete solution; i.e., the value of the discrete solution in the following sample time interval. The declaration of SAMPLETIME variables is as follows:

```
                                                             *
<sampletime declaration> ::= SAMPLETIME <samp id> {, <samp id>}
                                                             0
<samp id> ::= <id> ( { <number> } )
```

Notice, that if the sample time is varying, the length of the first sample time interval must be specified in the parentheses and the formula for updating the sample time should be given as a model equation (assignment statement). The declaration of variables of the discrete type holding the next discrete solution value is as follows:

```
<discrete declaration> ::=
                                                             *
                       DISCRETE <discrete id> {, <discrete id>}
                                                             0
                          / <id>     \
<discrete id> ::= <id> ( {            } ) ( <id> )
                          \ <number> /
```

The first parentheses must contain either a sample time variable or a number giving a constant sample time. The second parentheses must hold the name of the discrete solution variable associated to the discrete variable. Notice, that both the sample time variable and the solution variable must be declared before the declaration of the discrete variable.

Note, that variables of type DISCRETE <u>cannot</u> appear in expressions, they can only appear once as left hand side of an assignment statement.

Below we give a small example that shows two ways of solving the equation $y' = -y$, $y(0) = 1$, and $t \in [0,2]$. The main difference between them is that one of the methods gives a smooth errorcontrolled solution and the other gives a 'step solution'.

```
BEGIN
SAMPLETIME  H(0.1);           (*  CONSTANT STEPSIZE         *)
VARIABLE YGOOD(1), YBAD(1);(*  THE CONTINUOUS AND THE
                                  DISCRETE VARIABLE         *)
TIME T(0:2);
DISCRETE YNEW(H)(YBAD);       (*  THE 'NEXT' SOL.           *)
DERIVATIVE DY(YGOOD);
YNEW := YBAD + H*(-YBAD);     (*  EULERS METHOD             *)
DY := -YGOOD;                 (*  BUILT-IN METHOD           *)
WRITE(YGOOD, YBAD, YGOOD-YBAD)  (*  GENERATE OUTPUT         *)
END.
```

The use of difference equations should only be applied when the physical system behaves as a discrete system; that is, when the system has discontinuities - this is the case for example in a sampled control system. The facility should never be used as in the above example to solve differential equations using a bad method for the discretization.

6.2.2. Assignments.

Syntactically there are two types of assignments, but semantically there are more. SIL very much looks like any other programming language in the way assignment statements are constructed. The syntax is given below.

<assignment> ::= <lefthand side> := <expression>

<lefthand side> ::= { <id> | <number> }

We will in this section describe the statements whose left hand side is an <id>. These statements say that the variable on the left hand side is assigned a value, and the value is determined by evaluating the expression on the right hand side. Accordingly, all variables appearing in the expression on the right hand side must have been assigned a value either by declaration or by a previous assignment. Expressions are described in a later section.

As an example of the use of an explicit assignment of a value and a simple model we take the following problem:

$$y' = \lambda\ y, \quad \lambda = -10, \quad y(0)=1, \quad 0 < t < 1$$

A SIL program containing this model could be :

```
BEGIN
VARIABLE  Y(1);            (*  Y with initial value 1  *)
PARAMETER LAMBDA(-10);     (*  The lambda as PARAMETER *)
DERIVATIVE YDOT(Y);        (*  The DERIVATIVE of Y     *)
YDOT := LAMBDA*Y           (*  The equation            *)
END.
```

An equivalent program is :

```
BEGIN
VARIABLE  Y(1),  LAMBDA;
DERIVATIVE YDOT(Y);
LAMBDA := -10;
YDOT := LAMBDA*Y
END.
```

We have here a parametric use of the variable LAMBDA which is allowed since the compiler will determine that the statement

```
LAMBDA := -10
```

only need to be executed once - as an initialization. The computation of YDOT is done as often as the numerical method for solving the ordinary differential equation needs it.

This example also shows how it is possible to declare parametric variables and to use these as auxillary variables.

If a model contains a system of algebraic equations WITHOUT algebraic loops then it is possible to reformulate these as a sequence of assignments like those just shown. Because the statements of the model are (roughly) sorted, one is not allowed to assign the same variable twice.

Parameters, dynamic variables (variables with a derivative), and discrete variables normally cannot be assigned a value by an assignment statement. The only exception is when initial values are stated for dynamic variables or discrete variables declared with an empty parentheses '()' instead of the initial value. The initial value is supposed to be computable based on the right hand side of an assignment statement; the right hand side must therefore only depend on (some of) the declared parameters.

6.2.2.1. Expression.

One of the key elements of almost any programming language is the expression. In chapter 6.1 Fundamentals, we have given the syntax for it. At first, it looks rather simple but due to its recursive nature it can be used for any type of formula or expression. The recursion is primarily used to keep track of the nesting of parentheses and to allow an expression as argument to the standard functions. The syntax ensures the standard hierarchy among the operators +, -, *, /, and ** . The hierarchy assures that for example A*B + C*D is evaluated correctly.

The expression generally results in a floating point result. This is because all types of variables declared in a SIL model are stored as floating point numbers. The accuracy of a calculation depends on how floating point numbers are represented on the computer. It is a SIL philosophy to use at least 64 bit reals. This gives 14-16 correct digits in any intermediate calculation with a numeric range from approximately 10^{-100} to 10^{+100}. Normally, the SIL user does not have to worry about this, but the number of correct digits indicates that a relative tolerance of 10^{-10} is about the best one can expect from the numerical integrator. This is also of importance when dealing with implicit algebraic equations; stating a bound on the residual which is too small compared to the magnitude of the terms in the residual equation normally results in the error message saying that too many iterations have been taken (no solution found within the specified accuracy).

6.2.2.2. Discrete Variable Assignment.

The small example in section 6.2.1.7 DISCRETE and SAMPLETIME shows a simple use of discrete variables. These variables are only updated (evaluated) at the beginning of each of the time intervals in which they are kept constant. All other variables are evaluated each time the integrator requires it, and they are not evaluated at monotonic time points.

The update of a discrete variable is controlled by one sample time variable. If this facility is used for modelling a queue in which the arrival and departure times are different, each of the two processes must have their own sample time variable and their own discrete variable holding the number of arrivals and the number of departures, respectively. The queue size is then the difference of the two variables. Below we give an example of a problem where the sample time is varying with time.

```
BEGIN
SAMPLETIME STEP(0.1);             (* the sample time is def. *)
VARIABLE Y(1), DISY();            (* the discrete variable   *)
DISCRETE NEWY(STEP)(DISY);
DERIVATIVE DY(Y);
TIME T(0:20);
DISY := 5;                         (* the initial solution   *)
STEP := 1.1 - Y;                   (* determine sampletime   *)
NEWY := DISY + STEP*(-0.2*DISY);   (* difference equation    *)
DY := -0.5*Y;                      (* differential equation *)
WRITE(DISY,Y)
END.
```

The discrete solution variable DISY is declared with empty parentheses, indicating that we will give the initial value as a statement. The model also contains a statement that defines the desired sample time sequence; notice that the initial sample time is given in the declaration. In this example the sample time depends on the actual solution of the differential equation. The integration interval is [0,20].

6.2.2.3. Predeclared Variables and Functions.

As a special case of the assignment statement we have certain types of control statements. If the left hand side is one of the following reserved words the meaning is as described below. The right hand side must be an integer or a real number.

ABSERROR: The maximum permitted absolute value of the local error during the integration. The default value is 10^{-5}.

RELERROR: The maximum permitted relative value of the local error (relative to the components of the solution). The default value is 10^{-5}.

STEPSIZE: The initial integration stepsize. The default value (= 0) of STEPSIZE causes the integrator itself to determine the size of the first step. The (non-zero) value of STEPSIZE is also used as initial stepsize after the passage of a discontinuity.

MAXSTEPSIZE: The upper limit of the stepsize. Can be useful when the model has discontinuities to ensure that the code does not miss one. Its value is also used in connection with noise (random number generators). The random number is scaled by the ratio of the actual stepsize to MAXSTEPSIZE. The default value is no upper limit (0.0) unless the model uses NOISE.

INTEGRATOR: Choice of program for the integration. At the moment the underlined routiones are not implemented.

	INTERP.	ALGOLW	PASCAL	FORTRAN
= 1	STRIDE	STRIDE	STRIDE	STRIDE
= 2	SIRKMUS	SIRKMUS	SIRKMUS	SIMPLE

Other routines may be added in the future. In appendix D there is a documentation for the implemented routines. STRIDE is default.

LANGUAGE: Choice of language used for the simulation. The different implementations of SIL will have different options available for this variable. The list of possible choises is given below.

= 0: Interpreter: The model is interpreted each time it is evaluated; this is the default (and only) option in μSIL.

= 1: ALGOLW: For historic reasons ALGOLW was chosen as the first intermediate language to which SIL compiled the models. This language is on its way out and so is this option.

= 2: PASCAL: Since SIL is written in PASCAL this option is normally available on most systems. It requires however (contrary to the interpreter option) that a PASCAL compiler is present on the computer.

= 3: FORTRAN: The only reason for making the SIL compiler generate FORTRAN code is to be able to call some other generally available routines for integrating the system of ordinary differential equations. This option is not yet available.

METHOD: The integration subroutines/procedures often have a choice between several options. The value of this variable is used for that control. See the documentation of the specific integration procedures/subroutines in appendix D for further information. The default value is a non-stiff method.

MAXORDER: Some of the subroutines use variable order methods during the integration and this variable is used to give an upper limit on the order. Default value is 10.

MAXCPU: The maximum number of CPU-seconds that a simulation may execute. After each accepted step of the integration, SIL will check the CPU-time used so far. A value of 0 (zero) will turn the check off and this is the default.

MAXITER: When solving the (non-linear) system of coupled algebraic equations SIL will take MAXITER iterations if no solution satisfies the other stopping criteria imposed. This option serves as a safe-guard if the stopping criteria is too restrictive compared to the numerical accuracy used by the code. This option is also useful if (by accident) there is no solution to the system equations. The default value is 250.

The last six variables must have an integer as the right hand side.

Standard functions known in SIL include the standard mathematical functions SIN for sine, COS for cosine, ATAN for Arctangent, EXP for the exponential, LOG for the natural logarithm, and SQRT for square root. Notice, that the argument for SIN and COS is in radians and that SQRT is symmetric; that is, it accepts negative arguments. The function ABS for absolute value is missing because it has a discontinuity in its first derivative. Use an IF THEN ELSE statement to obtain the same effect. See the section 6.2.4 Branching for further details.

Besides the mathematical functions, SIL has a noise generator. The syntax allows the reserved word NOISE to be followed by a parentheses containing either an initial seed for the random number generator or a type specification of the noise or both. Both of these are given as integer numbers. For example, we can have the following assignment in a model

```
Y := ..... *NOISE(3295,1) + ....
```

saying that a random number generator of type 1 (see later) and with an initial seed of 3295 should be used when evaluating the variable Y.

Type = 1 corresponds to a uniform distribution of the random numbers in the interval [0,1];

type = 2 corresponds to a gaussian (normal) distribution with a mean value of 0 and a variance of 1; and

type = 3 will give a uniform distribution of the random numbers in [-1, 1].

For all three of these distributions the random numbers will be scaled with the ratio 'actual stepsize'/MAXSTEPSIZE; the corresponding unscaled distributions appear as type = 4, type = 5, and type = 6, respectively. Changing the seed value will change the sequence of random numbers but not their distribution.

The NOISE function requires that MAXSTEPSIZE is set properly. If not set by the user μSIL will set it to 1/20 of the integration interval. The reason for this requirement is that the random numbers are scaled with stepsize/MAXSTEPSIZE.

Other functions are planned for a future version of the SIL program, these include INTEGRAL and DELAY.

6.2.3. Implicit Assignment.

As mentioned earlier a model often consists of a system of ordinary differential equations coupled with a system of algebraic equations which can not be solved explicitly. In SIL the implicit equations are given by the implicit assignment statement.

If a <number> is on the left hand side of ':=' the assignment defines an algebraic equation which should be satisfied such that the residual is numerically less than the left hand side. If, in an implicit assignment statement, some of the variables have not previously been assigned a value, these variables become the unknowns in the system of algebraic equations and the number of these variables must be equal to the number of equations.

As a small example on this facility we give a program for solving two linear equations with two unknowns :

```
BEGIN
PARAMETER                         (*  The coefficients    *)
           A11(1),     A12(2),
           A21(3),     A22(4);
VARIABLE
           Y1,         Y2,        (*  The solution  *)
           B1,         B2;        (*  Right hand side  *)

B1 := 8;    B2 := 18;

1.0'-6 := B1 - (A11*Y1 + A12*Y2);
1.0'-6 := B2 - (A21*Y1 + A22*Y2);

WRITE(Y1, Y2)
END.
```

Notice that a small residual has nothing to do with the actual error of the computed solution; therefore a solution is both checked by SIL with a standard error test and with the test on the residual before it is accepted. The standard error test requires the change (DELTA) on any solution component Y to satisfy

$$\text{ABS(DELTA)} \leq \text{ABSERROR} + \text{RELERROR} * \text{ABS(Y)} \quad ,$$

where ABSERROR and RELERROR are the predeclared error tolerance variables described earlier.

SIL allows both implicit algebraic equations and implicit differential equations; that is, both variables and derivatives may appear as the unknowns in implicit equations. The only restriction is that in one implicit equation there cannot be both an implicit derivative and an implicit static variable as unknown. In that particular case SIL will assume the implicit static variable as a known quantity for which the value is determined by other equations.

Normally, only implicit static variables are adjusted during the iterative solution of the algebraic equations. In the following example TEMP is one of these special explicit static variables mentioned in section 6.2.1.3. VARIABLE; here it is used in connection with implicit derivatives and an IF THEN ELSE construction. This example also illustrates that the equation for an implicit variable or derivative in the ELSE part need not be an implicit assignment statement.

```
BEGIN
VARIABLE Y(1), TEMP();
DERIVATIVE DY(Y)(2.5); (* impl. deriv. with init. val *)
TIME T(0:2.5);
IF T < 1.8 THEN
   BEGIN
   TEMP := DY*DY - Y;
   1.0'-6 := TEMP - DY*(DY+1); (* effectively Y' = -Y *)
   END
ELSE
   BEGIN
   TEMP := 0;
   DY := 0.1             (* the derivative is constant *)
   END;
WRITE(Y)
END.
```

If it had not been for the empty parentheses following the declaration of TEMP, that variable would have been assumed dependent on the variable DY. Error messages would have been issued both in the first assignment of TEMP (since DY has not appeared in any statement) and in the implicit assignment statement (since TEMP depends on an implicit derivative).

This example also illustrates that if an implicit variable occurs in the THEN part of an IF THEN ELSE construction it must also appear in the ELSE part; implicit variables need not appear in an implicit equation in the ELSE part though, because SIL will itself rewrite the explicit equation to implicit form. This form can be used to specify the initial guess for the implicit variables/derivatives when the model changes from the ELSE state back to the THEN state.

6.2.4. Branching.

The language elements already described permit us to build models of various systems. All these models have in common that the state is unchanged throughout the integration. Often though, one has systems which switch between different states.

A shift of state normally causes a jump discontinuity in $\underline{f}$ and consequently in the derivative of $\underline{y}$. This gives a break in the solution curve $\underline{y}(t)$. The reason for the change of state can be an event occuring at a certain time, it can also be a condition depending on the solution, which is no longer fulfilled.

We will use simple branching in the description of such systems. Since, if the system is not in one state it must be in the other, and since a given state can "branch" into sub-states a model can have an arbitrary number of states.

The syntax of branching is :

```
<ifthenelse> ::= IF <switch expr> THEN <statement> ELSE
                                                 <statement>

                                                 *
<switch expr> ::= <switch term> { OR <switch term>}
                                                 0

                                                    *
<switch term> ::= <switch factor> {AND <switch factor>}
                                                    0

                         1  ⎧ <switch id>       ⎫
<switch factor> ::= {^}     ⎨ (<switch expr>)   ⎬
                         0  ⎩ <relation>        ⎭

                                 ⎧ > ⎫
<relation> ::= <expression>     ⎨   ⎬   <expression>
                                 ⎩ < ⎭
```

Note that in a <relation> the first expression must not start with '('. The compiler will detect the parenthesis as the start of a (<switch expr>) which obviously may be wrong. We can illustrate this by an example: The statement

```
IF ((A+B)*2 < C OR TEST) AND RELAY THEN ......
```

will cause an error message; a small change in the statement will avoid this problem

```
IF (2*(A+B) < C OR TEST) AND RELAY THEN ......
```

The problem could have been solved in other ways too.

The function ABS seems to be missing in the list of standard functions. This is intentional, because ABS would otherwise introduce an additional discontinuity in the model, and it would be a discontinuity which was not detected by the built-in algorithm. It could be very tempting to program the absolute value function Y := ABS(X) like Y := SQRT(X*X) but don't; the correct way to model the absolute value is like

```
IF X > 0 THEN
   Y := X
ELSE
   Y := -X;
```

This way is simply the fastest, specially if the argument X often passes through 0 (zero).

The <statement> clauses in the IF THEN ELSE statement will often be a <block>; that is, a group of statements surrounded by BEGIN and END. Notice, that there must be a statement following THEN and a statement following ELSE.

We also need the syntax for assignment of values to SWITCH variables.

```
<switch assignment> ::= <switch id> := <relation>

<switch id> ::= <id>
```

Note the similarity with assignment of other variables - the right hand side is just a relation between two expressions.

As an example of a simple switch in states we can take the following differential equation :

$$\left.\begin{array}{lll} y' = y & \text{for} & t < 1 \\ y' = -y & \text{for} & t > 1 \end{array}\right\} \quad 0 < t < 2, \quad y(0) = 1$$

This becomes in SIL :

```
BEGIN
VARIABLE Y(1);              (*  a variable   *)
DERIVATIVE YDOT(Y);         (*  its derivative *)
TIME T(0:2);

IF T < 1 THEN
   YDOT := Y                (*  one state    *)
ELSE
   YDOT := -Y;              (*  the other state *)

PRINT(Y)
END.
```

Here the <switch expr> is a <relation>. If we introduce a variable of type SWITCH the example could be formed as :

```
BEGIN
VARIABLE Y(1);        DERIVATIVE YDOT(Y);
TIME T(0:2);
SWITCH REL1;             (*  the SWITCH   *)

REL1 := T < 1;           (*  and the assignment of it  *)
IF REL1 THEN
   YDOT := Y
ELSE
   YDOT := -Y;

PRINT(Y)
END.
```

It is actually in this way that SIL always handles relations appearing between IF and THEN. A more advanced use of SWITCH variables is visualized in the following example. It is a very simple model of an ON/OFF temperature regulator.

$$
\begin{array}{llll}
(1) & y' = y & \text{while} & y < 2 \\
(2) & y' = -y & \text{while} & y > 1
\end{array}
\qquad 0 < t < 10, \quad y(0) = 1.5_1
$$

At first glance the model looks like a contradiction since both equations seem to hold when $1 < y < 2$. The model should be read in the following way: When the first equation is active it is active as long as $y < 2$. When y equals 2 the state is shifted and the other equation becomes active with 2 as the initial value. This equation will be active as long as $y > 1$. When y equals 1 we shift back to the first equation and so on. If the initial value is in the interval [1,2] where both equations may be active then we must indicate which equation should be active at the beginning. This is done here by the subscript 1 on the initial value 1.5. In SIL it is done by giving the switch a suitable initial value. A SIL model for solving this problem is given below.

```
BEGIN
VARIABLE Y(1);        DERIVATIVE YDOT(Y);
TIME T(0:2);
SWITCH REL2(ON);
IF REL2 THEN
   BEGIN
   YDOT := Y;
   REL2 := Y < 2          (*  stays ON until Y hits 2 *)
   END
ELSE
   BEGIN
   YDOT := -Y;
   REL2 := Y < 1          (*  stays OFF until Y hits 1 *)
   END;
PRINT(Y)
END.
```

Note that in the ELSE part the inequality is reversed. The logical explanation is that in order to reach the ELSE part, REL2 must be OFF (false) and accordingly the negated condition is used in the ELSE part.

The examples just described are relatively simple in their structure and the conditions for switching. It is possible though to build more complex switch expressions since AND and OR are legal operators in a SWITCH expression. They cannot appear in the assignment of a SWITCH variable, though. The correct use is illustrated in the following example :

```
BEGIN
VARIABLE Y(1);        DERIVATIVE YDOT(Y);
TIME T(0:2);
SWITCH REL2(ON);

IF T < 5 AND REL2 THEN      (* both must be ON (true) *)
   BEGIN
   YDOT := Y;
   REL2 := Y < 2
   END
ELSE
   BEGIN
   YDOT := -Y;
   REL2 := Y < 1
   END;

PRINT(Y)
END.
```

Each of the states (the statements after either THEN or ELSE) may be divided into several sub-states by nested IF THEN ELSE statements. This allows for a very complicated structure and using AND and OR in the switching condition it is possible to model rather complicated situations. Notice however, that the more complicated the structure and the conditions are, the more careful one has to be when proof-reading. Errors in the switching conditions may lead to very unexpected results.

When using an IF THEN ELSE statement one thing in particular must be observed. It is essential that all the equations in the THEN statement, say, can be evaluated even though the condition is false (OFF). The reason being that the numerical technique used for locating the shifting point requires that the model is solved <u>past</u> that point. Below we give an example of a model which is syntactically correct but the IF statement does not ensure that the LOG function is called only for positive values of X.

```
:
:
IF X > 0 THEN
   Y := LOG(X)
ELSE
   Y := -X;
:
```

6.2.5. Calling MACROs.

In order to be able to use MACROs as defined in the section 6.2.1.6 MACRO there is also a mechanism in the SIL language for calling (expanding) a previously defined macro. The syntax is as follows:

```
<macrocall> ::= <macro id> { ( <actual parm list> ) }^1_0

<macro id> ::= <id>

<actual parm list> ::= <actual parm> {, <actual parm>}^*_0

                    ( <id>     )
<actual parm> ::=  {  <number>  }
                    ( <int>    )
```

A macro is called/expanded whenever its name <macro id> is met in the SIL program. The actual parameters for the macro must be placed in the parentheses following the name. They must be given in the same order as the variables they replace in the macro and the actual parameters must match each of the formal parameters in type. <number> or <int> in the calling sequence are only allowed if the corresponding formal parameters are of type PARAMETER.

The macro is expanded immediately during compilation and the formal parameters in the macro model equations are replaced by their actual parameters. It is the users responsibility to ensure that the expansion leads to a valid model; for example that the expansion of a macro does not cause multiple assignments of a variable. SIL will catch these errors, but the error message will refer to the line number of the macro call. Below we give a small example of nested declaration and the use of MACROs.

```
BEGIN

MACRO OUTER(VARIABLE YO;
            DERIVATIVE DYO;
            PARAMETER G;
            TIME T);
   BEGIN
   MACRO INNER(VARIABLE YI;
               DERIVATIVE DYI;
               PARAMETER SIGN);
      DYI := SIGN * YI;           (*  INNER MACRO EQUATION *)
                          (*  OUTER MACRO EQUATIONS FOLLOW *)
   IF T < G THEN
      INNER(YO, DYO, 1)
   ELSE
      INNER(YO, DYO, -1)
   END;                                 (*  OF OUTER MACRO *)

TIME T(0:2);
VARIABLE Y(1);
DERIVATIVE DY(Y);

OUTER(Y, DY, 0.8, T);         (*  EXPANDING THE MACRO *)

WRITE(Y)                                    (*  OUTPUT *)
END.
```

This example could be written much easier without using macros. If this example is compiled with a $LIST line as one of the first lines in the file, the expanded macros will be listed in the .LST file associated with the model. This listing will show that the macro INNER is expanded into the macro OUTER twice, and the macro OUTER is expanded into the main model once. During these expansions some temporary variables will be generated but the final size (the number of equations) of the model will be (normally) no larger than if the statements have been expanded by hand.

6.2.6. Specification of Output.

One type of statement which always must be available in a SIL model is an output statement. No output statement in model ==> no results printed ==> no simulation performed. In the previous sections we have given some examples of output statements in connection with model examples. Here is the general syntax:

$$\langle\text{output statement}\rangle ::= \left\{ \begin{array}{l} \text{WRITE} \\ \text{PRINT} \\ \text{PLOT} \end{array} \right\} (\ \langle\text{output list}\rangle\ )$$

$$\langle\text{output list}\rangle ::= \{\ \langle\text{int}\rangle\ ,\}_0^1 \left\{ \begin{array}{l} \langle\text{expr}\rangle\ \{,\ \langle\text{expr}\rangle\}_0^4 \\ \langle\text{id}\rangle\ (\ \langle\text{id}\rangle\ ) \end{array} \right\}$$

The expression <expr> is a standard expression. The simplest expression is a variable. For output elements that are not variables SIL will generate an explicit variable and assign to it the value of the expression. The name of that variable will be artificially generated.

The three different output procedures generate different forms of output. The differences are mainly seen when running μSIL in batch mode (see section 3.1). In this case

WRITE	will give a table of the results in the .LST file,
PRINT	will produce a line printer plot of the results, and
PLOT	has the same effect as PRINT. It is intended to be used in connection with a plotting device.

During the solution process all output points will be sampled by the μSIL system. Each output statement gives rise to one sample series. If the first parameter in the parentheses is a positive integer number, the whole solution interval will be divided into that many subintervals. The variables specified will be sampled at each border of each subinterval. Including the samples at each endpoint, this means that the number of samples equals the number specified plus 1. If the first integer is omitted a default value will be used. The default is 55 for WRITE, 100 for PRINT, and 250 for PLOT. Below we give some examples of output statements.

```
  :
WRITE(100, A, B, EXP(C));
PRINT(X);
PRINT(1024, X(Y));
  :
```

For the PRINT and PLOT statements there are two ways of displaying an output variable. If the variables are given in a simple list with separating commas, μSIL will produce a time plot. In the batch version the Y-axes are scaled differently for each variable in order to obtain the best usage of the plotting area. In the graphic command system (see section 7.4) the scaling of the plotting area can be changed interactively but the solution is clipped at the plotting area boundary.

The other possible plotting mode is the so called phase plane plot. The user may specify that any variable should be plotted against any other variable. On the lineprinter plot the plotting area will be 101 times 101 characters. The number of sampled points can be important for the quality of the plot, since the points are not interconnected.

In any of the lineprinter plots the plotting character will be the first character of the name of the plotted variable. In case the name is artificially generated the first character will be the argument number. The multistrike character is '*'. If an error occurs during the solution process all of the output points sampled until then will be used to produce the 'up to now' results. If the model is a static model with no time variable the results can be written to the .LST file using the WRITE statement. The smallest possible running SIL program is given below.

```
WRITE(2+2) .
```

The number of variables that can appear in one output statement is limited to 5. The time variable always appears automatically in all WRITE statements. There could be a maximum number of 10 output statements in a model, giving a total of 50 output variables in any model. The number of samples in an output statement as well as the total number of samples may have an installation dependent maximum.

6.3. Semantic Rules and Consequences.

The syntax of the SIL language does not prohibit that the same variable is assigned a value more than once, but the semantics of the language does. In this chapter the semantic rules will be given and some of their consequences will be shown.

Rule no. 1.

A variable (or a derivative) must have been assigned a value before it can be used (in an expression) in the model.

A variable is assigned a value when it has appeared as the left hand side of an assignment statement. Parameters, dynamic variables, or implicit variables are assigned 'variables' either per definition or when given an initial value. In case a model contains what is often referred to as an algebraic loop, the model cannot be implemented using explicit assignment statements without breaking this rule; implicit assignment statements should be used in this situation.

Rule no. 2.

A variable (or a derivative) can only be assigned a value once; variables assigned in an IF-THEN-ELSE construction must be assigned in both the 'THEN' part and the 'ELSE' part.

This rule ensures that the sorting of the statements done by the μSIL system is safe. A variable will never appear in a model equation before it has been assigned a value, and in all parts of the model, the same value will be used. The second part of rule number 2 requires all variables (SWITCH variables being an exception) to have a well-defined value in all states of a model; this protects against 'undefined' variables being used in the calculation of subsequent equations.

Rule no. 3.

All variables must be declared before they are used.

It is possible in the SIL language to mix declaration statements and other statements as long as rule number 3 is not violated. It is good practice though, to put all the declarations at the top of the SIL-model. In some situations the SIL language compiler will generate additional variables; their names are constructed such that they cannot conflict the name of any of the variables defined in the model.

Rule no. 4.

At most one TIME variable can be declared.

A SIL simulation model only allows derivatives with respect to one model time (which need not be 'physical' time), therefore it does not make sense to define more than one model time.

Rule no. 5.

Variables of type SWITCH cannot appear in the right hand side of an assignment.

SWITCH variables are to be treated as boolean variables with the value being either ON or OFF. The value normally determines which one of two model states is active.

Rule no. 6.

Only variables of type VARIABLE or DERIVATIVE can be 'simple parameters' in the output statements; variables of type TIME or PARAMETER can only appear in 'output expressions'.

The output statements always contain the model time as an implicit first parameter, and since variables of type PARAMETER cannot change value during a simulation, neither of these variables are allowed for direct output.

6.4. Compiler Directives and Output.

During compilation the SIL compiler will echo the SIL model as read from the .SIL file into the .LST file. If errors are detected in the model the error messages will be printed in the .LST file. When solving the model in 'batch' mode the file contains also the results from the simulation as well as some run statistics. This chapter describes the basic structure of the .LST file and the various compiler directives available to the user.

The .LST file is designed to be printed on a 132 character printer with 60 lines pr. page. Many PC printers have though a condensed output format; with this the .LST file will fit on a page 80 characters wide.

An example of a .LST file can be found in fig. 3.2; with reference to that figure we can give the following details. A page used for echoing the model has a 2 lines page header. The first line contains the version number of the μSIL system and the date of its release.

The second line has in its first 11 positions the user descripter code compiled into the μSIL system; this code is unique for each copy of the system. The second field on the line is the date when this copy was generated. The third field is a 40 characters wide running title set by the $TITLE compiler directive (see later). The fourth field is the date and time for generating this file and the last field is the running page number.

The header is separated from the listing of the model by two blank lines. A line of model listing starts with the line number followed by a 2 characters field which is either '--' or has one or both of the '-' overwritten with a digit. Corresponding BEGINs and ENDs have the same digit overwriting the first, respectively the last, '-'. The rest of the line is exactly as read from the .SIL file.

At most 55 lines of a model is printed on one page; the compiler automatically generates a formfeed (FF) and a new page header before the 56th line is printed, and this line becomes the first line of the new page.

If the compiler does not detect any errors, the last page of model printing will contain also a line giving the total compile time in seconds and some lines summarizing the number of variables, derivatives, parameters etc.. If errors were found during the compilation the messages will be printed on separate pages following the complete printing of the model.

When running a correct model in batch mode the following may be found on the subsequent pages:

1) If for some reason (too much CPU-time used, range check on a variable failed, simulation breaked by the user (the 's'-key), etc.) the simulation is stopped prematurely, the error messages is printed together with a complete variable dump.
2) Statistics collected during the solution phase will be printed together with a list of options used.
3) All parameters and their values are printed followed by a list of all dynamic variables and their initial values and their final values.

4) For each output statement in the model there will be one page (at least) of the corresponding generated output.
5) The .LST file is always terminated with two lines giving the total execution time and the amount of memory left on the computer (this indicates how much bigger a model that can be solved).

When running the model in graphic mode the .LST file normally contains only the printing of the model and the amount of memory left on the computer. The printing of the model will be missing whenever the editor has been invoked.

There are 4 different compiler directives available. A directive requires a special line starting with a '$' in column 1. Such lines will never be part of the model and they will not appear in the listing of the model in the .LST file. The 4 current directives are list, nolist, title, and debug. They have the following keywords 'LIST' , 'NOLIST' , 'TITLE=' , and 'DEBUG,' .

A directive line suppressing printing of the model is as follows,

```
$NOLIST
```

with '$' being in column 1. The printing of the model will then be disabled until the line

```
$LIST
```

is met in the source file (.SIL). The list and nolist directives actually works on a level basis; that is, after, say, 3 nolists printing is first resumed after 3 list directives. Initially, the compiler starts printing on level 1, level 0 turns off printing, whereas level 2 activates the printing of MACROs when they are expanded. Fig 6.1 contains the listing of a macro model. The use of the list directive at level 2 causes the inclusion of the expansion of the MACROs.

```
SIL VERSION 2.0 (880808)
TEST VERS.  86-10-27


    1 1-  BEGIN
    2 --  (*    first macro    *)
    3 --  MACRO FIRST(PARAMETER A;
    4 --              VARIABLE P;
    5 --              DERIVATIVE PDOT);
    6 2-    BEGIN
    7 --    (*    the innermost macro    *)
    8 --    MACRO INNER(VARIABLE S,SQ);
    9 3-      BEGIN
   10 --      IF S > 0 THEN
   11 --         SQ := SQRT(S)
   12 --      ELSE
   13 --         SQ := SQRT(-S)
   14 -3      END;
   15 --
   16 --    VARIABLE TEMP, SQTEMP;
   17 --    TEMP := A*P;

EXPANSION OF MACRO INNER
********************************************
BEGIN
  IF TEMP > 0 THEN
  SQTEMP := SQRT ( TEMP )
  ELSE
  SQTEMP := SQRT ( - TEMP ) END ;

********************************************
END OF MACRO INNER
   18 --    INNER (TEMP, SQTEMP);
   19 --    PDOT := -TEMP + LOG(SQTEMP + 10)
   20 -2    END;
   21 --
   22 --  VARIABLE Y1(1), Y2(5), Y3(-3);
   23 --  DERIVATIVE DY1(Y1), DY2(Y2), DY3(Y3);
   24 --
   25 --  TIME T(0:2);
   26 --
EXPANSION OF MACRO FIRST
********************************************
BEGIN
  VARIABLE TEMP , SQTEMP ;
  TEMP := A * Y1 ;
  BEGIN
    IF TEMP > 0 THEN
    SQTEMP := SQRT ( TEMP )
    ELSE
    SQTEMP := SQRT ( - TEMP ) END ;
  DY2 := - TEMP + LOG ( SQTEMP + 10 ) END ;

********************************************
END OF MACRO FIRST
   27 --  FIRST(10, Y1, DY2);
```

```
EXPANSION OF MACRO FIRST
********************************************
BEGIN
  VARIABLE TEMP , SQTEMP ;
  TEMP := A * Y2 ;
  BEGIN
    IF TEMP > 0 THEN
    SQTEMP := SQRT ( TEMP )
    ELSE
    SQTEMP := SQRT ( - TEMP ) END ;
  DY3 := - TEMP + LOG ( SQTEMP + 10 ) END ;

********************************************
END OF MACRO FIRST
  28 --   FIRST(-1.5, Y2, DY3);

EXPANSION OF MACRO FIRST
********************************************
BEGIN
  VARIABLE TEMP , SQTEMP ;
  TEMP := A * Y3 ;
  BEGIN
    IF TEMP > 0 THEN
    SQTEMP := SQRT ( TEMP )
    ELSE
    SQTEMP := SQRT ( - TEMP ) END ;
  DY1 := - TEMP + LOG ( SQTEMP + 10 ) END ;

********************************************
END OF MACRO FIRST
  29 --   FIRST(0.11, Y3, DY1);
  30 --   WRITE(Y1, Y2, Y3)
  31 -1   END.

  3.25  SECONDS IN COMPILATION

MODEL CONSISTS OF :
    3 PARAMETERS
    6 EXPLICIT STATIC VARIABLES
    3 DYNAMIC VARIABLES / DERIVATIVES
    3 SWITCH VARIABLES
```

Figure 6.1: Listing of a SIL model with MACROs; the expansions are printed because the list level is 2.

The title directive is used for changing the running title of the listing in the second line of the page header. Also, this directive forces skipping to the next page (with the new title). This forced pageskip is not activated if the directive is the first line of the .SIL file. The 40 characters immediately following the '=' will be the new title (blanks are added if the line is not 47 characters long). The title from the last title directive in the model file will be used in the graphic command system as the main title above the graphical drawing area. Below we show an example of a title directive

```
$TITLE=This is a new running title
```

The debug directive can be used both when debugging a model and when debugging the whole μSIL simulation system. The following line

```
$DEBUG,3
```

sets the debug level to 3. Initially, the debug level is 1. Debug levels from 0 to 5 affects the generation of intermediate output during the solution phase of the simulation and debug levels from 5 to 8 causes the compiler to generate information concerning its internal operation. Debug level 9 is a 'print all' option.

The directive has effect from the following line and until the option is changed by a new directive. The last directive in a model will be the one used during the whole solution phase (it cannot be changed in that phase). Increasing the debug level increases the amount of output generated, and normally a user should never use the debug levels from 5 to 9. All the debug output will appear in the .LST file.

Since the debug directive has effect on the whole solution phase independent of whether it is activated in batch or from the graphic command system it should be used with care. Each SIM command will add the debug output to the .LST file. Also, the use of the debugging facility considerably decreases the efficiency of the solve process, and debugging large models will produce an enormous amount of output.

7. The μSIL System.

In this chapter we shall describe the μSIL system in details. The SIL language is used for defining the model which is to be simulated by the system. We will now see how such a model can be entered into the μSIL system, how it can be manipulated (changed, extended or reduced), how it is compiled, how it is run, and how the results are presented.

The compiler and especially the SIL language was described in chapter 6. The main sections of this chapter are 7.1 The editor, 7.2 The command processor, and 7.3 The graphics. This completes the description of the μSIL system.

The SIL language is designed to cope with possibly almost any type of simulation model (except those resulting in partial differential equations). Future developments in the language will be incorporation of new facilities for 'new' problems. One object of the language is also to force the user to structurize his/her model; it is necessary for the user to have a much clearer understanding of what the mathematical model of the system is, which states he has and what the conditions for switching from one state to another are.

When building SIL models we strongly recommend that the first time a model is typed, it is done with an ordinary text editor. There are many good reasons for this. The first one is that the user normally has a favorite editor; the second reason is that the editor in the μSIL system is line-oriented but full-screen editors are much more efficient to work with; the third reason is that it is impossible to save the edited file on disk unless it is a correct SIL model. For the rest of this chapter we assume that a SIL model (or a first approximation to it) is available in a file called MODEL.SIL .

When running the μSIL system there are two modes of operation. It can be run either in batch mode or interactively. **In batch** mode the SIL model is compiled and solved without giving the user any other interrupt possibilities other than to stop the solution phase. The results can only be presented as either tables or as line printer plots in the file MODEL.LST.

In **interactive** mode the user may alternate bewteen editing the model and solving the model without leaving the μSIL system. The editor has online syntax control. The solution process is part of a command driven graphic output processing system which can use most graphics display boards available for PC's. The command mode is designed to make it easy to 'play' with a model (change parameter values and initial values) in order to investigate the characteristics of the model and get information about its behaviour.

7.1. System Overview.

There are six major components in the μSIL system. These are:

1) the SIL language compiler,
2) the interactive line editor,
3) the numerical integration routine (STRIDE),
4) the model interpreter,
5) the interactive command processor, and
6) the graphic output processor.

The activation of each of these components is decided in a small main program. In fig. 7.1 we show the logical interconnection between the various components of the system and the input and output files used.

From the figure we can mention the following characteristics.

1) The input file (MODEL.SIL) is read either directly by the compiler (syntax analyser) or it is read by the editor. The editor passes the model to the compiler.
2) The compiler produces a listing of the model in the output file (MODEL.LST); when compilation is finished control is relinquished to the command processor via the main program.
3) In batch mode the command processor is bypassed and control is immediately given to the simulator. When the problem is solved, the graphic component processes the output and puts the results into the output file.

4) In non-batch modes the command processor is initiated; it initializes the graphics system, it splits the screen into 4 sub-areas, and draws a coordinate system in the graphic output area. It then waits for a command from the user.
5) When the SIM command is given the command processor activates the simulator for solving the problem. During the solution process the solution is simultaneously drawn in the graphic output area. When the solution is finished control returns to the command processor.
6) When a solution is sampled, the command processor can activate the graphic output system to display it directly.

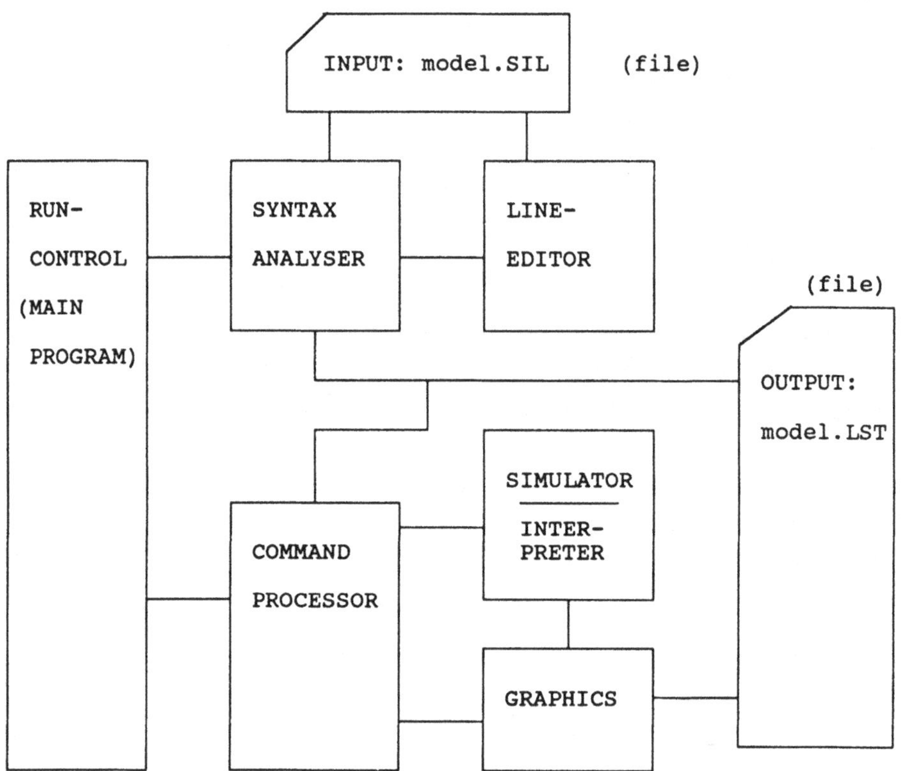

Figure 7.1: Schematic structure of the μSIL system.

In batch mode the different parts of the μSIL system are not recognized by the user; it works as a one-shot (compile go) system. In the interactive mode the user has to be concerned about the editor and its commands, and the command processor.

7.1.1. μSIL System Files and Surroundings.

This section is concerned with how the μSIL system is installed on the PC, which files are required and what are they used for; it also contains a description of the different ways that the μSIL system can be invoked.

The μSIL system as it is normally distributed consists primarily of files matching the 'wildcard' names SIL.* and *.BGI . The *.BGI files contain the BORLAND graphic device drivers for the different types of graphic screens that μSIL supports. The μSIL system itself is in the file SIL.EXE , the error messages and the online help information are in the files SIL.ERM and SIL.HLP , respectively, and SIL.ACC contains some vital accounting information. These files are all essential for a proper function of the μSIL system and they must all be in the same directory; as shown later, the name of that directory is not essential.

Figure 7.2 shows a diagram of a fictitious directory tree. The three important directories are

1) the current directory,
2) the μSIL system directory, and
3) the directories containing SIL models.

The situation schetched is rather general; normally, the current directory and the directory for SIL models will be the same. In a multi drive system none of these three directories need to be on the same disk drive.

Before we describe how the μSIL system can be used in this situation, let us have a look at which files the different directories must and may hold.

The directory \SYSTEM\MATH\SIL contains the whole μSIL system. This is all the files from the distribution diskette matching the 'wildcard' names SIL*.* and *.BGI . The SIL model directories contain the model files; they must all match the 'wildcard' name *.SIL .

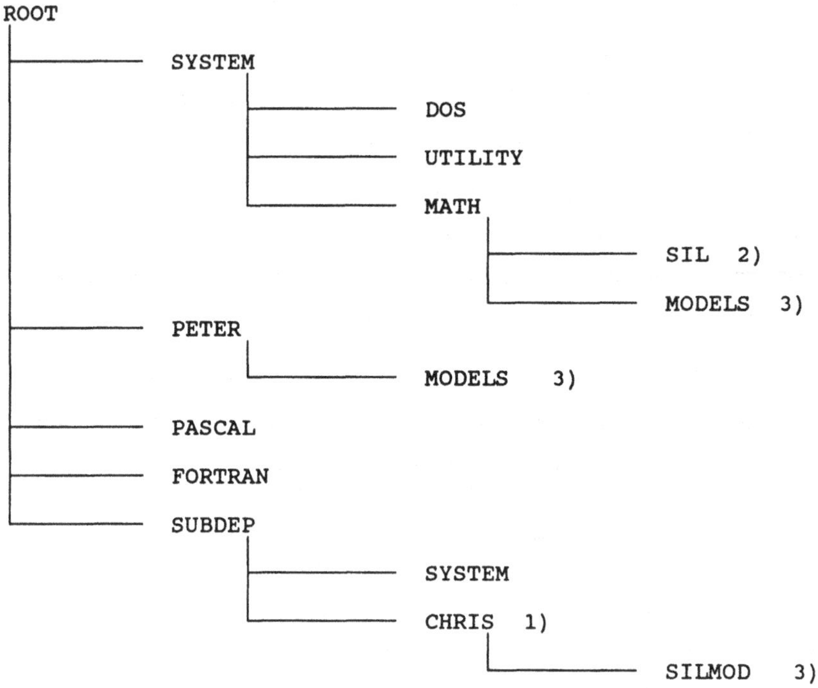

Figure 7.2: Fictitious directory tree with 1) = current directory, 2) = directory for the μSIL system, and 3) = directories containing SIL models.

The current directory \SUBDEP\CHRIS should contain the file SIL.DEV . The first line of this file must be the complete path to the directory containing the μSIL system files; that is, where can the μSIL system locate the files SIL.ACC , SIL.ERM , and SIL.HLP . With the above structure of the directory tree, the SIL.DEV file should contain the following line

```
C:\SYSTEM\MATH\SIL
```

if our fictitious directory is on drive C. If on invocation, μSIL is unable to locate the file SIL.DEV in the current directory, it will search the directory tree on the current drive for the file SIL.EXE . If it is found, the SIL.DEV file is automatically generated in the current directory. The SIL.DEV file must be in all directories from where μSIL is activated.

Using the same notation as in chapter 6 we can give the general form of the command used to invoke the μSIL system. This is

```
               1      1               1          1
{{<drive>:} <path>} SIL_{<name>{.SIL} {_(<option>}
               0      0               0          0
```

The underline '_' denotes that one or more blanks must be inserted. Elsewhere, blanks are not accepted. <drive> indicates the disk drive on which the μSIL system resides and <path> is the full DOS path to the directory in which the μSIL system is located; if the DOS command PATH=<SIL-path> has been issued, both <drive> and <path> may be omitted. <name> is the file name of the file containing the SIL model; <name> may include both a <drive> and a <path> to the directory where the file is located. The file must have the file extension SIL (following the period); this extension may or may not be given explicitly.

The <u>first character</u> following the left parenthesis '(' is used as <option>; recognized characters are 'g', 'G', 'i', 'I', 'b', and 'B'. 'G' stands for Graphics and with this option the μSIL system will take you directly into the graphic command system after having compiled the model. 'I' stands for Interactive, and μSIL will invoke the editor on the file specified. 'B' stands for Batch, and with this option (or any other character than those shown here) μSIL will compile and run the model in one step.

Suppose Chris wishes to run his own model (MODEL.SIL) in the subdirectory SILMOD . He can activate μSIL with the command

```
SIL SILMOD\MODEL
```

or if there is no PATH to the directory where μSIL is stored

```
C:\SYSTEM\MATH\SIL\SIL SILMOD\MODEL
```

The results of the simulation will be written into the file MODEL.LST in the directory \SUBDEP\CHRIS\SILMOD .

Simulation of the models in the directory \PETER\MODELS can be executed by Chris using the following command (for interactive invocation):

```
\SYSTEM\MATH\SIL\SIL \PETER\MODELS\MODEL.SIL  (I
```

The listing from the compilation will be stored in a file with the full file name \PETER\MODELS\MODEL.LST , the results, of course, will appear on the graphic display.

The simplest situation is when the SIL directory 2) is also the current directory 1) and it also contains the SIL model 3). In this case we can use the command

```
SIL MODEL (g
```

if we want to immediately enter the µSIL system in graphic mode. The same simplicity is acheived provided 1) and 3) are the same directory and a PATH command like

```
PATH=\SYSTEM\MATH\SIL;
```

has been issued.

7.2. The Editor.

The editor in the µSIL system is a line editor. It is not intended for the primary creation of the model - use a good fullscreen text editor for that. The editor is intended to be used for making minor changes in a model; adding or deleting a term in an equation, permanently changing initial values or parameter values and the like without having to leave the µSIL system. It can also be advantageously used when compiling the model for the first time; typographical errors are easily corrected.

The editor operates with an actual or a current line. That line is always displayed prior to the command line prompt '=>' which indicates that the editor is waiting for a command to be entered. Commands are entered on a line by line basis; several commands can be entered on the same command line provided they are seperated by a ';' semicolon. Commands can be given in either upper case, lower case, or mixed; in the rest of this chapter commands are always given in upper case for clarity. A command need not be typed in full length; the editor will recognize any abbreviation of the commands but in this manual all commands are written in full length. For some of the commands the editor actually accepts some typing errors; TIP will, for example, be interpreted as the command TOP. A command line can be edited using the DOS editing keys (Backspace and left arrow <-) until the line is terminated by ENTER (Carriage Return).

Each line of the model has a line number. The line number is always relative to the start of the model file. Inserting new lines or deleting lines will change the line numbers for the following lines. The line number is always displayed when the line is displayed. Two virtual lines are added to each model, one at the beginning and one at the end; they are called the 'TOP of file' and the 'BOTTOM of file' lines.

The first time the editor is activated in a μSIL 'session' it will read the MODEL.SIL file from disk into main memory. The copy in memory will be used for later invocations of the editor as long one does not leave the μSIL system. Notice that when the editor has been invoked there will be less storage available for the internal copy of the model and output points. On the first invocation of the editor, the user will be notified with the message

```
SIL, interactive editor
HELP is available
*** TOP of file
=> _
```

On later invocations only the *** TOP of file line will be displayed.

The editor commands fall in three categories: 1) positioning commands, 2) line changing commands, and 3) environment commands. For each category we will systematically explain all the commands.

Positioning commands.

The simplest positioning commands are TOP and BOT, respectively. TOP will position the editor at the virtual line before the first line in the model, and BOT will position the editor at the virtual line after the last line in the model. If the editor compile mode is ON (see later) the BOT command will only reach the bottom line if there are no errors in the model. If there are errors in the model, the editor will display the latest compiled line when the first error is encountered. The cause of the error may be found in a previous line.

Positioning at other lines in the model can be done by the JUMP command. This command positions the editor relative to the current line. Moving the current line 5 lines forward is done with the command JUMP +5 . This can be abbreviated to +5 or just 5 . Similarly, -3 positions the current line 3 lines backward. If the editor compile mode is ON (as is the default) moving backwards can be expensive because the JUMP command in this situation is interpreted as a TOP command followed by an appropriate forward JUMP . The TOP command will re-initiate the compiler and the model will be re-compiled to the new current line.

The LIST command also repositions the editor. LIST +5 is the same as JUMP +5 with the exception that all the lines from and including the old current line up to and including the new current line (5 lines ahead) will be listed on the display. The LIST command can also be used like LIST 5,10 , meaning that the lines between number 5 and 10 are to be displayed. 5 and 10 are here absolute line numbers and line number 10 becomes the new current line.

With the COMPILE command specific lines can be compiled (and checked for errors). COMPILE 5 will compile the next 5 lines and the last one will become the current line. If the editor compile mode is OFF the model is recompiled up to the current line before the 5 lines are compiled. The compilation either stops when the 5'th line (relative) is reached or when the first error is encountered (which could be before the current line).

The FIND command is used to search for a specific string in the model. The search is always initiated in the line following the current line. A match is found when the line currently investigated contains a string which is precisely the same as the search string; upper and lower case as well as the number of blanks must match. If the FIND command is given without a parameter, the search string is either the search string from the previous FIND command or the first replacement string from the previous REPLACE command (see later). FIND /abc/ will search for the string 'abc', as will the command FI4abc4 . The first non-blank character that is not part of the command is used as string delimiter; end-of-line (ENTER) also terminates the string.

Line modifying commands.

The group of line modifying commands consists of DELETE, INPUT, NEW, and REPLACE. They are described below.

With the DELETE command one line or several lines can be deleted from the model. There is no recovery facility in the editor, so use this command with care. DELETE without parameters will remove the current line, making the following line current. DELETE +3 will remove the current line plus the 2 (two) following lines. DELETE -3 removes the current line plus the 2 (two) most preceding lines; the line before those becomes the new current line. DELETE 6,9 will remove the lines 6, 7, 8, and 9. Line 10 becomes the new current line and its line number will be changed to 6 because the line numbers are always updated when lines are deleted or new lines are inserted.

Inserting new lines after the current line can be done using the INPUT command. When activated without parameters it creates a new empty line and inserts it immediately after the current line; the editor is then switched to input mode, where everything entered on the command line is written into this new line. An ENTER (Carriage Return) will make this line the current line and the editor will continue writing on a new empty line. Entering an additional empty line (2 ENTERs following each other) terminates the INPUT command. INPUT +3 works as INPUT with the exception that it automatically terminates when 3 new lines have been inserted.

Rewriting lines can be done using the command NEW. Without parameters NEW will display the current line and prompt for a new content of the line. The new line must be terminated with ENTER (Carriage Return). An empty line (a line with no characters except an ENTER) will restore the original content of the line. The empty line also terminates the command and this line becomes the current line. NEW+3 will prompt for a new content of the current line and the 2 (two) following lines one at a time. The last line treated by the command becomes the current line.

Another way of changing the content of a line is by using the REPLACE command. This command can substitute a string of characters in the current line with another string. The strings in the command line are case sensitive; that is, 'a' and 'A' are different. The command

REPLACE/abc/ABC/ will change the <u>first</u> occurrence of the string 'abc' in the current line to the string 'ABC'. The first non-blank character following REPLACE will be used as delimiter; that is its next occurrence in the command line will separate the search string from the replace string. REPLACE abcaBCa therefore exchanges 'bc' with 'BC'. If not terminated in other ways, the replace string is terminated by the end of the command line. REPLACE without a parameter will use the search string and the replace string from the most previous REPLACE command; this allows sequences of F;R;F;R;F;R to be used for substituting one string for another through the whole model (only once pr. line). Empty strings are accepted.

Environment commands.

The environment commands are used for controlling the operation of the editor. These commands include ABORT, END, SET, and HELP. They are described below.

The editing session can be terminated in one of two ways. The command ABORT immediately terminates both the editor and the whole μSIL system. This is the way to leave the editor without saving any of the changes made in the model file so far. This is the only command that can transfer control from the editor directly to DOS . The END command also terminates the editor, provided the SIL model is syntactically correct. When the model is correctly compiled by the μSIL compiler, the model is stored back on disk, and the interactive graphic command system is invoked automatically. If the model contains an error detected by the compiler, the END command will not be completed. The last compiled line will become the current line and the error message will be displayed, and nothing is stored on disk. The editor can only store (save) models which are syntactically correct.

The editor has two controls accessible by the user. The first control is called compile mode, and it determines whether or not the editor keeps all of the model above the current line compiled. The second control is list mode; when it is off, most commands display only the current line after their completion. When it is on, some commands will display the lines as they are 'worked on'. The SET command is used to change any of the two controls. SET COMPILE OFF will turn off the compile mode; this will make the editor faster in some situations (in compile mode the REPLACE command actually causes a recompilation of the model above the current line). It will not prevent the compilation performed by the END command. SET COMPILE ON turns on the compile mode again and this is also the default setting. The command SET LIST ON turns on the list mode. In list mode a command like COMPILE 16,25 will also list lines 16 to 25 on the screen as they are compiled. The default is SET LIST OFF in which case only the current line is displayed on the screen after the completion of a command.

The HELP command provides on-line help for all editor commands as well as all the commands in the graphic command system. HELP without a parameter will display all the valid editor commands. HELP followed by a command name will display on the screen any information from the SIL.HLP file concerning that subject. Any editor command may be arbitrarily shortened, whereas HELP for commands from the graphic command system requires the command to be fully typed.

The rest of this chapter is an alphabetic list of all the editor commands with a description similar to the one in the HELP system.

Function: ABORT

Parameters: No parameters.

Examples: ABORT

Description: This command immediately terminates the edit session without updating the model on the disk. It is mainly for use if everything seems to go wrong. See also END.

Function: BOTTOM

Parameters: No parameters.

Examples: BOTTOM; Bot; B

Description: The line pointer is moved to a virtual line following the last line of the file. This virtual line then becomes the actual line. In COMPILE mode, the rest of the model will be analyzed; if any errors are found, the editor will stop moving and display the line containing the error (or perhaps the following line).

Function: COMPILE

Parameters: <number> | +<number> | -<number>

Examples: COMPILE +5; C-3; Comp 5

Description: The number of lines specified are compiled by SIL. If NOT in COMPILE mode, all the preceding lines are also compiled without any notice to the user other than error messages, if necessary. The last line becomes the actual line. If the compiler is in LIST mode the lines compiled will be printed on the terminal.

Function: DELETE

Parameters: <number> | +<number> | -<number> |
<number>,<number> | ,<number> | <number>,

Examples: DELETE 5; dEL-3; D10,15; DELE,32

Description: The lines in the specified range are deleted and the line preceeding the deleted lines becomes the actual line. The lines are effectively deleted from the model and there is no recovery facility, therefore use this command with care. If the range contains a ',' (comma) the line numbers are absolute, otherwise they are relative to the current actual line. If the command appears in a one line sequence of commands, absolute line numbers will be those valid before any command is executed.

Function: END

Parameters: No parameters.

Examples: END; e; En

Description: Terminates the edit session. The model is first analyzed by SIL. If the model is correct, it is rewritten to disk and the model is run by the processing system. If the model is not correct, SIL will display the line with the error (or the following line) as the actual line and terminate the END command without storing anything; thus, it is impossible to leave the editor via the END command with a syntactically incorrect model. See ABORT.

Function: FIND

Parameter: <string>

Examples: FIND /abc/; Fi7abc7; f.abc.; F

Description: Starting with the line following the actual line, the editor will search the model for the first location of the string parameter. The first non-blank character that is not part of the command is used as string delimiter. The search string is case sensitive; that is, upper and lower case characters in the search string do matter. If no string is specified, then the string used in the last FIND or REPLACE command will be used. See REPLACE.

Function: HELP

Parameters: <command>

Examples: HELP; He END; HELP input

Description: Gives online assistance. The command displays parts of the file SIL.HLP on the screen. This file contains most of the descriptions given here for each command. HELP without a parameter it displays a list of all commands available plus some editor status information.

Function: INPUT

Parameters: <number>

Examples: INPUT; In 3; I; I17

Description: This command puts the editor in INPUT mode; that is, subsequent lines are inserted in the file just after the actual line. This mode is exited for one of three reasons.....

1: an empty line (not a blank line) is entered

2: if in COMPILE mode an erroneous statement is typed

3: if INPUT <nr> , maximum <nr> lines can be inputted.

Note: In the command INPUT<nr> , <nr> can be any positive number and blanks may be inserted. An empty command (ENTER) is equvalent to INPUT.

Function: JUMP

Parameters: <number> | +<number> | -<number>

Examples: JUMP 3; J+2; -3; +5; 22

Description: This command is used to position the actual line in the model. Moving is always relative to the current actual line. + is forward and - is backward. JUMP without a parameter is equivalent to +1 or just 1. The editor will make the next line of the model the actual line. For convenience, JUMP+5 is the same as +5 or just 5. Movement to an absolute line number can be done by the combined command TOP;+5 , this makes line number 5 the actual line.

Function: LIST

Parameters: <number> | +<number> |
<number>,<number> | ,<number> | <number>,

Examples: LIST+7; l 3; L10,17

Description: The lines in the specified range are listed on the display terminal and the last one becomes the actual line. The range may be specified as a relative line number (relative to the actual line) or it may be specified as an absolute line number interval. If the editor is in COMPILE mode the lines are simultaneously analyzed by SIL.

Function: NEW

Parameters: <number> | +<number> | -<number> | <number>,<number> | ,<number> | <number>,

Examples: NEW+3; n10,15; New ,22; nEW -6

Description: The lines in the specified range will be typed on the screen one by one and the user is prompted for new replacement lines. An empty line (not a blank line) will terminate the command immediately. In the COMPILE mode a syntax error will also terminate this command. If the range specified contains ',' (comma) the line numbers are assumed to be absolute line numbers. A missing line number is assumed to be the number of the actual line.

Function: REPLACE

Parameters: <string1> <string2>

Examples: REPLACE /abc/ABC/; Rep4abc4ABC4; R;

Description: In the actual line, the first location of <string1> will be exchanged with <string2>. If no parameters are specified, then the parameters from the latest FIND or REPLACE command will be used. All characters except semicolon can be used as delimiter. Either <string1> or <string2> may be the empty string. Issuing a FIND command without a search string immediately after a REPLACE will use <string1> as search string. A REPLACE without parameters after a FIND command with parameter will erase the string.

Function: SET

Parameters: <option> ON | <option> OFF

Examples: SET COMPILE ON; S C OFF;
SET L OFF; S LIST ON

Description: Used for selecting between different modes of the compiler. Valid commands are:
SET COMPILE ON (Default)
Causes the editor to make a SIL analysis of everything above and including the actual line.
SET COMPILE OFF
The model is not automatically analysed.
SET LIST ON
Forces the editor to list on the terminal the lines treated by commands like COMPILE and JUMP .
SET LIST OFF (Default)
Types only the actual line on the terminal immediately after a command or a sequence of commands on one line.

The keywords COMPILE and LIST may be abbreviated down to C and L respectively; ON and OFF must be fully typed.

Function: TOP

Parameters: No parameters.

Examples: TOP; T; toP

Description: This command will cause the line pointer (the actual line) to be moved to a virtual line before the first line of the model.

7.3. The Command Processor.

This chapter describes the command processor of the interactive graphic command system. When necessary, parts of the graphic system are described but a complete description of the graphic system can be found in chapter 7.4 The graphics.

The command processor operates with four different fields for the communication with the user.

1) The command input field; in this area all commands typed by the user will be displayed character by character. The prompt '>' here indicates that μSIL is waiting for the user to type a command.
2) The graphic output field; this is the largest area on the screen and is used by μSIL for presenting graphically the results as they are generated.
3) The status field; in this 3 line area, the μSIL system will display status information (the number of integration steps and the model time) during the solution process. If a run error stops the solution the relevant error number is also displayed in this area.
4) The text output field; used for displaying the value of some (or all) of the variables and used for displaying the HELP information.

Normally, μSIL operates in a mode where all four fields are displayed together on one graphic page but it also has a mode where the input and the output fields are displayed on a text page, and the graphic field and the status field are displayed on a graphic page. The switching between the two pages occurs automatically. The rest of this chapter mainly concerns the commands that the user can enter in the input field and the consequences it may have in some of the other fields.

Whenever the μSIL system is ready to accept a command line for the command processor it will prompt in the input field with a '>'. Commands are read line by line; this means that the user may enter several commands on the same line (up to 75 characters long), a semicolon ';' must separate the different commands. The command line will be executed by the command processor when it is terminated by ENTER (Carriage Return). Editing the command line is only possible via the Backspace key (deletes the last character). Contrary to the editor, all commands must be typed in full length, but can be in upper or lower case, or mixed.

The commands can be divided into 3 groups; they are

1) interaction with the model:
   PLOT, SHOW, SIM, and changing parameter values, initial values, or predefined system variables.
2) operations concerning the graphics:
   DRAW, TITLE, XAXIS, and YAXIS.
3) environment commands (and others):
   CLEAR, EDIT, END, FULLSCREEN, GRAPHIC, HELP, and DUMP_VARS.

As for the editor, we will describe the various commands in the order of appearence in the different groups.

Interaction with the model.

A SIL model always contains at least one output statement (WRITE, PRINT, or PLOT). The last output statement in a model will by default be the one used for selecting the variables drawn during the solution process. The PLOT command can be used to add one extra (virtual) output statement to the actual model. Additional use of the PLOT command just overwrites the previous virtual output statement. The syntax of the PLOT command is precisely the same as the PLOT output statement in the SIL language with one exception: Only variables can be specified in the output list, not expressions. The variables specified must of course be defined in the SIL model.

The SHOW command can be used for displaying the values of different kinds of variables and obtaining status information (see later). SHOW Y or just Y will display the actual value of the variable Y (if it is declared in the model) in the output field. If Y is a dynamic variable the initial value is also displayed. SHOW without parameter will display all variables in the model and their values, all system variables and their values, and the status of the output statements. The different groups of variables can be SHOWed individually, like

SHOW PARAMETER, SHOW SYSTEM, or SHOW SAMPLE. SHOW accepts VARIABLE, DERIVATIVE, STATIC, PARAMETER, SWITCH, SYSTEM, SAMPLE and any defined variable as a parameter.

The model is never solved by the μSIL system unless it is told to actually do so. This is done by means of the SIM command. When issued without a parameter the SIM command will cause the model to be solved in the time interval either as used in the most preceeding SIM command or as specified in the SIL model; the first one having preference to the latter. The most recent initial values for the dynamic variables will be used. The time interval for integration can be specified as parameter to the command; SIM(0.1:1.3E3) being an example. The time interval is given as lower bound and upper bound, respectively, as two real numbers separated with colon ':'.

If the lower bound coincides with the time at which the most preceeding integration stopped, integration continues from that point with the final solution as new initial solution. If for some reason the integration has been prematurely stopped before the end of the integration interval, it can be resumed from the point of interruption by the command SIM > . The SIM command in itself will not clear the graphic display area but the execution of other commands (CLEAR, DRAW, FULLSCREEN, PLOT, TITLE, XAXIS, and YAXIS) prior to the SIM command will force the erasure of the drawing area.

The value of parameters defined in the SIL model can be changed interactively. Also the initial values for either the defined dynamic variables, discrete variables, or implicit static variables can be changed. The predefined system variables are treated more or less as parameters and may be changed in the same way as well. The assignment is done using the assignment statement

<id> := <value>

where <id> is the name of the variable (or parameter) to be changed and <value> is its new (real or integer) value. If the initial value of a dynamic variable depends on a parameter (see sections 6.2.1.3 and 6.2.2), changing the value of the parameter also changes the initial value.

Suppose a model contains (among others) a parameter S and two dynamic variables Y1 and Y2. The initial value of Y1 is given in the model whereas the initial value of Y2 depends on the parameter S. The following command line

```
S := 0.2; Y1 := 5.3
```

(blanks do not matter) will change the value of S (and therefore implicitly also the initial value of Y2) and the initial value of Y1. Assigning Y2 explicitly, like

```
Y2 := -33
```

will remove the dependence of Y2 from S until the statement

```
Y2 := ;
```

reestablishes it again.

Operations concerning the graphics.

The commands for manipulating the graphical display can perform the following operations: Scale and label the x- and y-axis, change the title of the plot, or redraw any (sampled) variable either versus time or versus any other (sampled) variable without doing any simulation.

When a simulation is completed (using the SIM command) all variables sampled can be redrawn with the DRAW command or added to current display using the ADD command. A variable is sampled when it appears in either one of the WRITE, PRINT, or PLOT statements of the model, or in the last PLOT statement issued in the graphic command system. If a variable is sampled more than once, the most detailed sampling will be used.

DRAW and ADD are very similar: they accept the same argument lists, and they perform the same operations; the only difference being that the DRAW command clears the graphic drawing area. Both commands must be followed by a parenthesis in which the variables to be drawn or added appear. The command DRAW(A, B, C) will draw the variables A, B, and C versus model time. It is required that A, B, and C are defined in the model and sampled. The drawing area is cleared but the scaling of the axes is not changed. Similarly, ADD(A(B)) will draw the variable A versus the variable B with the current scaling of the axes without clearing the drawing area (not even if the scaling has been changed). If A and B are not sampled at the same points in time, both variables are automatically interpolated to match the sample points of the other variable. The DRAW command will always cause the following SIM command to clear the graphic drawing area but neither DRAW nor ADD will affect which variables are plotted during the SIM command.

At the top of the graphic drawing area there is a line serving as title of the plot. This line consists of two parts, a fixed part which identifies the μSIL system (or the user) and a part which is user defined. The command TITLE(New Title) will rewrite the user defined part of the title with the string 'New Title'. Note, the right parenthesis ')' serves as terminator of the string; it cannot itself appear in the title string.

The commands XAXIS and YAXIS are used for changing the scaling and the labelling of the x- and y-axis, respectively. Setting the range of the y-axis from -1 to 1 is done by the command YAXIS(-1:1). The command XAXIS(0:20.5,Time) will set the range of the x-axis from 0 to 20.5 and change/add the label 'Time'. The label of the y-axis is changed to 'Temp.' by the command YAXIS(,Temp.). There is no restriction on the lower bound being smaller than the upper bound on the axes. A right parenthesis cannot appear in the axis labels because ')' is used to terminate the string. Both of these two commands will immediately cause the appropriate axis to be redrawn with the new scaling. A SIM command following either of the commands will always clear the graphic drawing area before the new solution is drawn. The axes should be scaled by the user before any drawing is made (SIM, ADD or DRAW) and it is the users responsibility to ensure that the scalings are appropriate. Initially, the system tries to scale the axes dependent on the initial values; having been into the editor and returning back to the graphic command system normally preserves the scaling of the axes. Parts of the solution outside the range of the axes may not be visible, clipping is enforced.

Graphic environment commands.

These commands control various functions of the graphic command system. Normally, these functions are seldom used but that does not make them less important. There are commands for some control of the graphics in the system, a command (HELP) for online assistance, commands for terminating the graphic command system, and a command for writing all results to a file.

The CLEAR command has the effect that the following SIM command initially clears the graphic drawing area before a new solution is plotted. The effect is as if, say, the command YAXIS has been issued.

The EDIT and the END commands both terminate the graphic command system. END also terminates the μSIL system whereas EDIT invokes the interactive editor on the actual model. It should be noted, that when returning to the graphic command system after having edited the model, the scaling of the axis, the labels and the title, and the setting of the graphics normally are restored to the values obtained during the most previous graphic command session.

The μSIL system can operate its graphics in either of two modes. The mode initially selected has all the four fields, input field, output field, status field, and graphic display field on the screen simultaneously. In order to exploit the whole screen for the graphics, the other mode operates with one screen (in graphic mode) consisting of the graphic display field and the status field, and the other screen (in text mode) used for the input and the output fields. The SIM and the DRAW commands automatically switch to the graphic screen. When drawing is finished any keystroke will switch back to the text screen. The FULLSCREEN command toggles between these two modes of operation.

It is difficult always to select the most appropriate mode of operation of the graphic drivers used by the μSIL system. In this area of application we prefer high resolution graphics to many colours but sometimes this is not possible due to the hardware actually on the computer. By means of the command GRAPHIC the user may himself/herself select the most appropriate mode for the graphic drivers. The command will display all the possibilities and ask for a selection. It is not checked that the selected device is compatible with the graphics board on the computer, this is the users responsibility. A wrong choice may lead to unpredictable results: black screen, blinking screen, funny characters etc. Normally, a new call to GRAPHIC can be performed (even if the command is not echoed on the screen) and the appropriate choice made.

The HELP command in the graphic command system works very similar to the HELP command in the editor. HELP without a parameter displays all available commands. HELP followed by any of these commands will display whatever information is available on this command in the help file. HELP followed by any of the editor commands (fully typed) will display the information available on that command. HELP followed by any other name will search the help file for an entry and display the information found. HELP followed by a number (1-70) will search the file with error messages and display the error message with that number. This is helpful since run errors in FULLSCREEN mode are only reported in the status field using the number code.

The DUMP_VARS command will scan through all the output points currently sampled and append output processed as if all model output statements where WRITE statements to the <model>.LST file (where batch results normally appear). This is useful in the situation where the graphics produced by the μSIL system for one reason or another is not satisfactory. The .LST file generated by several calls to DUMP_VARS can be quit large.

Chapter 8. Examples contains several screen dumps which illustrate the use of most of the commands of the graphic command system.

The rest of this chapter contains an alphabetic description of the commands of the graphic command system organized in a structured manner. The explanations originate from the help file.

Function: ADD

Parameters: (<argument list>)
<argument list> :
1) <ident> , <ident> , ...
2) <ident> (<ident>)

Examples: ADD(A,B,C); Add(c(b)); add(B (a))

Description: With this function one can view the results of the preceeding simulation. All sampled variables can be plotted on the screen. The <argument list> takes the same form as in PLOT, except that <number> cannot be given. If a variable appears in several output statements SIL will always use the sample with most points. The ADD function will not clear the screen before the drawing is done. Note: the ADD command does not redefine the variables to be plotted by the next SIM command.

Function: CLEAR

Parameters: No parameters

Examples: CLEAR; clear

Description: After this command the following SIM command will always clear the graphic screen before the new solution is displayed. This clear option is automatically invoked by the DRAW, PLOT, XAXIS, and YAXIS commands and is always active in 'fullscreen' mode.

Function: DRAW

Parameters: (<argument list>)
<argument list> :
1) <ident> , <ident> , ...
2) <ident> (<ident>)

Examples: DRAW(A,B,C); Draw(c(b)); draw(B (a))

Description: With this function one can view the results of the preceeding simulation. All variables sampled can be plotted on the screen. The <argument list> takes the same form as in PLOT, except that <number> cannot be given. If a variable appears in several output statements SIL will always use the sample with most points. The DRAW function will clear the screen before the drawing is done, and furthermore, the screen will be erased by the next SIM command. Note: the DRAW command does not redefine the variables to be plotted by the next SIM command.

Function: DUMP_VARS

Parameters: No parameters

Examples: DUMP_VARS; Dump_vars

Description: This command will immediately print the values of all sampled variables in the .LST file. The format will be as if the model had used WRITE statements for all the variables.

Function: EDIT

Parameters: No parameters.

Examples: EDIT; edIT

Description: This command will terminate the graphic command system and start editing the current model. If SIL has been invoked with the (Interactive parameter the system will go into edit mode initially. See also the END command.

Function: END

Parameters: No parameters.

Examples: END; end

Description: The END command terminates the SIL system. The EDIT command will terminate the graphic command system but invoke the built-in interactive editor.

Function: FULLSCREEN

Parameters: No parameters

Examples: FULLSCREEN; fullscreen

Description: This command toggles the screen between a full page graphic screen with a separate input/output screen and a screen having both the graphic and the input/output fields.

Function: GRAPHIC

Parameters: No parameters

Examples: GRAPHIC; graphIC

Description: This command can be used for the setting of the graphic device drivers used by the μSIL system. The response to the command is a list of available device drivers and their different allowed modes; after this the current setting is displayed. The user is then requested to give the new setting. For multi page graphic boards also the active page can be selected.

Function: PLOT

Parameters: (<argument list>)
<argument list> :
1) {<number>,} <ident> , <ident> , ...
2) {<number>,} <ident> (<ident>)

Examples: PLOT(255,AB,CA,MYMAX); Plot(dy(y))

Description: This function defines new variables to be plotted during the following SIM commands. The <number>, argument is optional; if omitted, the default is 250. The <argument list> argument defines the variables to be sampled and plotted in the next simulation. The variables (up to 5) must be separated with a comma. The name of the variables must be defined in the SIL model; no distinction is made between lower and upper case characters. When the argument is <id1>(<id2>) a phase plane plot of <id1> versus <id2> is made. Notice, that a PLOT command will overwrite the information from a previous PLOT command (if any).

Function: SHOW

Parameters: <ident>
1) VARIABLE, DERIVATIVE, STATIC, PARAMETER, SWITCH, SYSTEM, SAMPLE
2) <variable identifier>

Examples: SHOW; Show system; SHOW STATIc; Show A; A, B, c

Description: This command will display the current values of all variables defined in the model. For dynamic variables the initial value is also displayed. The group of static variables (say) can have their values displayed by SHOW STATIC. SHOWing a variable will display the value of that variable at the current simulation time (SHOW Y is normally written as Y). This form of the command can be extended to <ident>, <ident>, <ident> ... .

Function: SIM

Parameters: (<number> : <number>)
>

Examples: SIM(0:10); sim; Sim(0.3:0.47); SIM >

Description: Performs the simulation. Without a parameter the default value for the integration interval (either [0,1] or as specified in the SIL program) will be used. If the left point of the specified integration interval coincides with the right point of the previous specified integration interval, the solution at that point is used as initial value. The SIM command **will not** clear the screen; by changing only initial values or parameter values all the results can be on the same graph. If a SIM command is interrupted (S-key), the solution may be continued with the command SIM > .

Function: TITLE

Parameters: (<string>)

Examples: TITLE(My plot); Title(Plotting A vs B)

Description: This function is used for entering a new title for the plot. Initially the title is 'SIL plot of results' unless the SIL model contains a title. The <string> may contain any character except ')' or carriage return. A maximum of 40 characters is allowed.

Function: XAXIS

Parameters: (<argument list>)
<argument list> :
1) <number> : <number> , <string>
2) <number> : <number>
3) , <string>

Examples: XAXIS(-10:10,Time); xaxis(2'-3:-4.1E-3);
XAxis(,X-label)

Description: This function changes the characteristics of the x-axis. The argument <value> : <value> will set the range of the axis as specified. The axis is redrawn with new divisions; the drawing area will not be cleared until the next DRAW or SIM command. The <string> parameter is used for labelling the axis; note, the first ')' will be the end of the string! A maximum of 10 characters is allowed.

Function: YAXIS

Parameters: (<argument list>)

<argument list> :

1) <number> : <number> , <string>

2) <number> : <number>

3) , <string>

Examples: YAXIS(-10:10,Solution); yaxis(2'-3:-4.1E-3);
YAxis(,Y-label)

Description: This function changes the characteristics of the y-axis. The argument <value> : <value> will set the range of the axis as specified. The axis is redrawn with new divisions; the drawing area will not be cleared until the next DRAW or SIM command. The <string> parameter is used for labelling the axis; note, the first ')' will be the end of the string! A maximum of 10 characters is allowed.

7.4. The Graphics.

This chapter is a general description of the graphic system implemented in μSIL. The graphics are designed primarily with respect to the user's needs when 'experimenting' with the model; it is easy to change parameters and initial values and immediately find the effects on the solution. Normally, the model time is plotted vertically (in the following referred to as the X-axis); in some situations though, the use of various phase plane plots is useful. In the μSIL graphics system the user may after one simulation (solution) plot any sampled variable versus any other (sampled) variable. There is no 'ZOOM' command but the user can choose any scaling of the drawing area he or she wants and thereby obtain any desired enlargement.

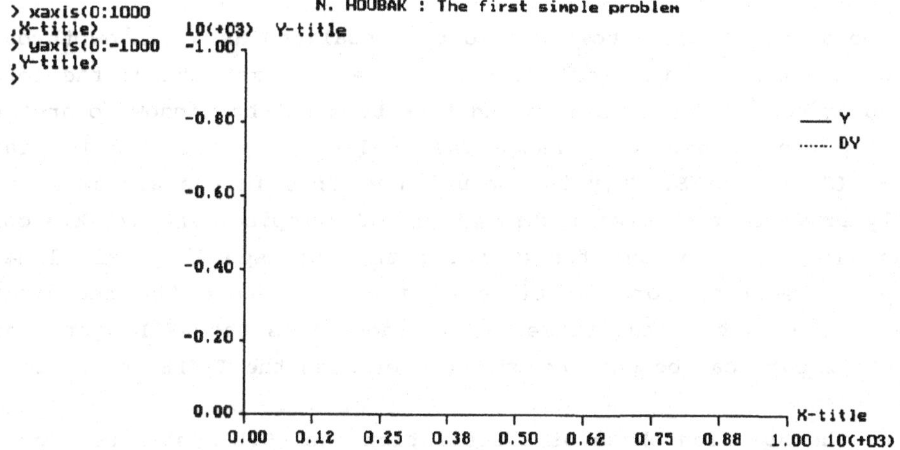

<u>Figure 7.3</u>: An example of how the graphic drawing area may look.

As mentioned in previous chapters μSIL has 4 input/output fields; a command input field, a graphic output field, a status field, and a text output field. In this chapter we consider mainly the graphic output field. Fig. 7.3 shows an example on this part of the graphic screen. A closer look at the figure reveals the following:

1) a title of the plot,
2) two coordinate axes with units, scaling, and label,
3) a drawing area for plotting the solutions, and
4) a field for displaying the names of the variables plotted and the line types used.

The position of the fields is fixed relative to one another.

The coordinate system always has the X-axis horizontal, below the drawing area and the Y-axis vertical, to the left of the drawing area. Normally, the X-axis represents the model time (a Time-axis) and is positive to the right; the Y-axis is positive upwards but scaling one of the axes with a lower bound bigger than the upper bound reverses the positive direction. By default, the X-axis is divided into 8 subintervals and the Y-axis is divided into 5. Each division is marked and the corresponding value is written below/to the left of the axis. The values are scaled automatically to 'nice numbers' and the scaling factor (powers of 10) is shown at the end of the axis. Also at the end of the axis, the user may add a label (a string of characters) indicating the units of the axis, the interpretation of the numbers, etc.

The coordinate system shown should be visualized as a window through which one looks at the real 2D plane. The user may change the lower and upper bounds for both axes and thereby move the window to another position and if desired, change the scaling. The commands for this are XAXIS and YAXIS. Only the solution visible in the window is actually drawn on the screen; this is called clipping. The X/YAXIS commands also have options for changing the corresponding axis label. Space is reserved for the title of the plot above the coordinate system; one part of the title always identifies the μSIL system and the other part can be changed by the user with the TITLE command.

During the solution phase of a simulation the μSIL system will generate a representation of the solution for selected variables. This generated representation is a table containing values of the independent variable (usually the time) and the corresponding values of the dependent (solution) variables. Normally, this sampling of the solution is done at time-equidistant points. One table can contain values of up to 5 solution variables and the system allows up to 10 such tables. The number of entries in a table has an installation dependent maximum. Also the total number of entries is limited (depends on the amount of storage available). If a simulation is prematurely terminated, the solution stored so far in the tables will be accessible for normal post processing. A variable having its values stored in a table is called a sampled variable. The same variable may appear in several tables (with different time spacing).

Each WRITE, PRINT, or PLOT statement in the SIL model generates one table. Each table may have its own time spacing. The time spacing is determined on basis of the number of entries in the table (a default or user defined) and the length of the integration interval. In addition to those tables, the user may in the graphic command system generate one extra table using the PLOT command. This is illustrated in fig. 7.4. Consequtive PLOT commands overwrite the last output table.

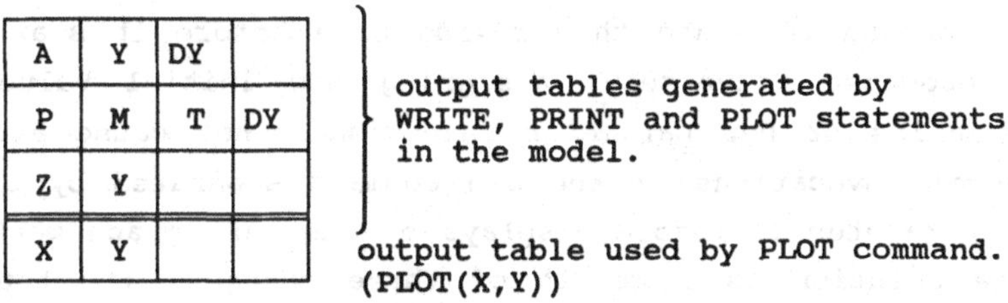

A	Y	DY	
P	M	T	DY
Z	Y		
X	Y		

output tables generated by WRITE, PRINT and PLOT statements in the model. (rows A, P, Z)

output table used by PLOT command. (PLOT(X,Y)) (row X)

Figure 7.4: Schematic description of the output table system.

During the solution process μSIL will simultaneously draw the solution generated for all variables in the last output table. This table is either the table generated by the most recent call of the PLOT command in the graphic command system or the table corresponding to the last output statement in the SIL model. The names of the variables drawn appear to the right of the drawing area with an indication of the type of line used for drawing the different variables. The solution of a variable is normally drawn by connecting the points in the table with straight lines. In order to make the solution smoother, the drawing generated during the SIM command also includes the solution obtained at the step points of the integrator (see chapter 5).

When the solution process is terminated (for one reason or another) the results are in the output tables, of which only the last one will be displayed on the graphic screen. The DRAW command can draw (in the window defined) any sampled variable versus time or versus any other sampled variable independent of whether the two variables are sampled at the same time points; it automatically interpolates both tables to match the time points in the other. The DRAW command will <u>not</u> auto scale the axes to make the whole solution fit into the window.

Plotting one variable versus another is referred to as a phase plane plot; it is a design goal in the µSIL system that any (changeable) variable can be sampled during the solution process, and any sampled variable can be plotted versus any other sampled variable. The smoothness of the phase plane curve DRAWn is primarily determined by the spacing of the output points. Phase plane plots generated during the SIM command are generally resonably smooth independent of the spacing of output points.

The commands DRAW, PLOT, TITLE, XAXIS, and YAXIS will cause the following SIM command to erase the drawing area before it starts the simulation process. Conversely, assigning new initial values for dynamic variables or new parameter values will not cause a screen clear. Several invocations of the SIM command separated by changes of, say, a parameter therefore displays on the same graph what happens to the solution as a result of these changes. It should be noted that only the solution from the last SIM command is available in the output tables for the DRAW command. The user may force an erasure of the drawing area with the CLEAR command.

The SIM command can terminate abnormally for several reasons: too much CPU-time has been used, a variable exceeds its validity range, too many iterations are requested, the integration steps are too small, a division by zero occurs, the argument of the LOG function becomes negative, or the user interrupts integration by pressing 's' or 'S'. If the problem can be resolved by changing a parameter or variable value, simulation can be continued by the command SIM > . It should be noted, that though the solution continues, it is a new invocation of the SIM command; this means that the output tables will only contain results from the last invocation of the SIM command. Previous values are not available.

In the initial default graphic display mode of the µSIL system the graphic display area occupies only 2/3 of the total screen area. The remaining 1/3 of the area is used by the various status, input, and output fields. The μSIL system has however a mode in which the graphic display area fills the whole screen, and the system then switches to a separate text screen whenever the input or output fields are needed. This mode is specially useful when the user wants to fully exploit the resolution of the graphic screen and perhaps make hard copies of the results produced. Since the graphic screen is erased between calls to the SIM command in this mode, it is impossible to have solutions from several SIM commands on the same graph. The command FULLSCREEN toggles the system between the two graphic display modes.

A hard copy of all the output tables can be produced by the DUMP_VARS command. This command appends to the .LST file (where all compiler output resides) output produced as if all the output tables where generated by WRITE statements in the model. Calling DUMP_VARS between several calls to the SIM command can create a rather large .LST file since it will contain a listing of all the output tables after each simulation. The μSIL system does not yet have a facility for making a hard copy of the screen; we refer the interested user to special software packages available on the market.

Until now the graphics has been treated without attention to which kind of graphics display board is installed in the computer, and that is how it should be - the type of graphics hardware does not matter. On the other hand, one cannot neglect the graphics hardware when it comes to making the graphics. The μSIL system uses the graphic primitives supplied with the TURBO PASCAL compiler used for compiling the system. This allows for automatic detection of the graphic hardware installed on the computer (without looking at the label on the display terminal). The graphic display boards supported by the compiler, and hence the μSIL system, are:

1) CGA, 2) MCGA, 3) EGA, 4) EGA64, 5) EGAMONO,
7) HERCULES, 8) ATT400 (or Olivetti), 9) VGA, and
10) PC3270.

The automatic detection of the hardware may fail to detect precisely the graphics on your computer or to initialize it properly; in this situation the GRAPHIC command for manually setting up the graphic software to a specific graphics board and a specific mode can be used. Some types of graphics boards have several pages (numbered 0, 1, 2,..); software used for making screen dumps on a printer, say, often need to know on which page the screen graphics reside.

8. Examples.

This chapter contains some of the models that SIL has been used to solve. Some of the examples are here merely for illustrating a specific facility of the system whereas other examples are real world simulation models. Some of them come from lecture notes while others come from research projects.

Example 1: The μbubble.

As the first example in this manual we will give a SIL model generated on basis of the very first real world model the author of these notes analyzed, built a numerical integration routine for, and solved. The results were published by Lewin and Bjørnø (1982).

The physical system is a gaseous microbubble in biological tissue, and we will investigate its response (variation of radius) to ultrasonic irradiation. The mathematical model is given in Lewin and Bjørnø (1982); it is a 2'nd order ordinary differential equation with a forcing term from the ultrasonic pressure field. The equation is given below.

$$\frac{d^2R}{dt^2} = \left[P_i\left(\frac{R_0}{R}\right)^{3\gamma} - P_0 - \frac{2\sigma}{R} - \frac{4\eta_\ell}{R}\frac{dR}{dt} - p(t) - \frac{3}{2}\rho_\ell\left(\frac{dR}{dt}\right)^2\right]/(\rho_\ell R)$$

where

$$P_i = \frac{2\sigma}{R_0} + P_0 \quad .$$

The dynamic variable is R, and it has the derivative dR/dt and the second derivative $d^2R/(dt)^2$. R_0 is the radius at equilibrium and the initial value for R. dR/dt is initialy set to 0 (zero). In metric units the parameters in the model have the following values:

$P_0 = 10^5$ Pa, $\sigma = 0.06$ N/m, $\rho_\ell = 1056$ kg/m^3

$R_0 = 2\cdot 10^{-6}$ m, $\gamma = 1.33$, $\eta_\ell = 0.003$ Pa·S

The forcing function p(t) is in the first part given by $p(t) = P_A \sin(f_A 2\pi\ t)$, where $P_A = 1.6\cdot 10^5$ Pa is the amplitude of the oscillating function and $f_A = 1.6$ MHz is the driving frequency.

```
$TITLE= Micro bubble (CE)
BEGIN

(*  Definition of model parameters   *)
PARAMETER  GAM(1.33), R0(2E-6), PINF(1.0E5),
           DEL(0.06),  AMP(1.6E5),  FREQ(1.6E6),
           PI(3.14159),  NY(0.003),    RHO(1056);

(*  Dyn. var. R  has an absolute error bound = 10^-7  *)
VARIABLE  R()(< 1.0E-7), REV_R,
          PT, PG, SIGNAL, OMEGA,  GAM3;

(*  Define first and second derivative   *)
DERIVATIVE  RDOT(R)(0), D2R(RDOT);

(*  Define model time   *)
TIME T(0:3.0E-6);

(*  Set integration parameters   *)
ABSERROR := 1.0E-10; RELERROR := 1.0E-3;

R := R0;       (*   initial value  *)

PG := 2*DEL/R0 + PINF;    (*  temporaries     *)
OMEGA := 2*PI * FREQ;     (*  and auxillary  *)
GAM3 := 3*GAM;            (*  equations       *)
REV_R := 1/R;

SIGNAL := AMP * SIN( OMEGA * T );
PT := SIGNAL;             (*  external pressure  *)

(*  Define the differential equation   *)
D2R := (PG * EXP(GAM3*LOG(REV_R*R0)) - PINF -
        (2*DEL + 4*NY*RDOT)*REV_R   -
           1.5*RHO*RDOT*RDOT      + PT) / (RHO*R);

(*  Output specification   *)
PRINT(250,SIGNAL);
PRINT(250,RDOT,PT,D2R);
PRINT(250,R)

END.
```

Figure 8.1: SIL model of the μbubble.

Fig 8.1 shows a SIL model of the equation; notice the following details:

1) the model is for continuous excitation (CE),
2) R_0 is used as the initial value for R,
3) R has it own absolute error tolerance (10^{-7}),
4) temporary variables are used for intermediate results
5) exponentiation with a real exponent is done using EXP and LOG.

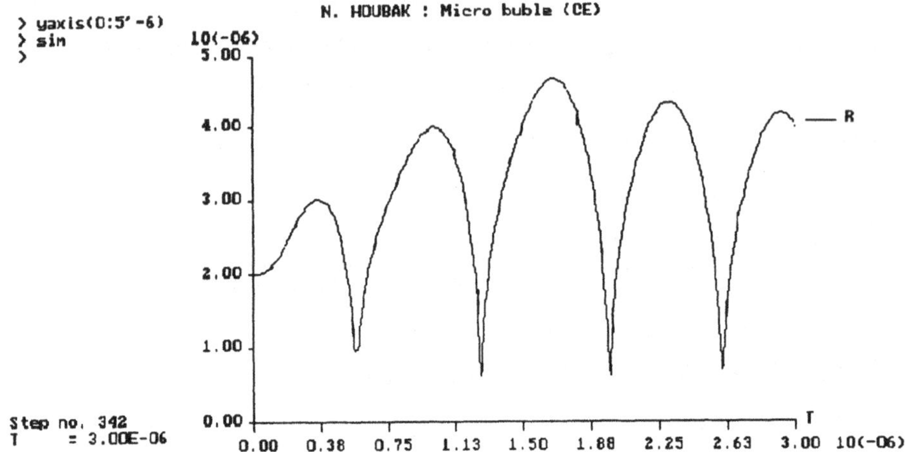

Figure 8.2: Graphic screen dump when solving the μbubble model.

The results of the simulation were obtained after approximately 2 minutes of computing and they are shown in fig. 8.2.

The following SIL model (fig. 8.3) shows the same model but now with a pulse excitation. The pulse is modeled by generating an envelope curve and then multiply with the signal.

```
$TITLE= Micro bubble (PE)
BEGIN

(*    Declarations    *)
PARAMETER  GAM(1.33), R0(2E-6), PINF(1.0E5),
           DEL(0.06),  TAU(3E-7),  TAU1(3E-6),
           AMP(1.6E5),  FREQ(1.6E6),
           PI(3.14159),  NY(0.003), RHO(1056);

VARIABLE  R()(< 1.0E-7), REV_R,
          PT, PG, ENVEL, SIGNAL,  ALFA,
          OMEGA,  ANGEL, GAM3;

DERIVATIVE  RDOT(R)(0), D2R(RDOT);

TIME T(0:5.0E-6);
ABSERROR := 1.0E-10;      RELERROR := 1.0E-3;
```

```
(*   Auxillary variables and initial values   *)
R := R0;                      REV_R := 1/R;
PG := 2*DEL/R0 + PINF;
GAM3 := 3*GAM;
ALFA := 1/TAU;

OMEGA := 2*PI * FREQ;     ANGEL := PI/(2*TAU);

(*  Generate the envelope curve    *)
IF T < TAU THEN
   ENVEL := 1 - COS( ANGEL * T )
ELSE
   IF T < TAU1 THEN
      ENVEL := 1
   ELSE
      ENVEL := EXP(-ALFA*(T-TAU1));
SIGNAL := AMP * SIN( OMEGA * T );
PT := SIGNAL*ENVEL;

(*    The differential equation     *)
D2R := (PG * EXP(GAM3*LOG(REV_R*R0)) - PINF -
           (2*DEL + 4*NY*RDOT)*REV_R   -
              1.5*RHO*RDOT*RDOT     + PT) / (RHO*R);

(*   Output specification    *)
PRINT(250,PT,ENVEL, SIGNAL);
PRINT(250,RDOT,D2R);
PRINT(250,R)
END.
```

Figure 8.3: SIL model of the μbubble with pulse excitation.

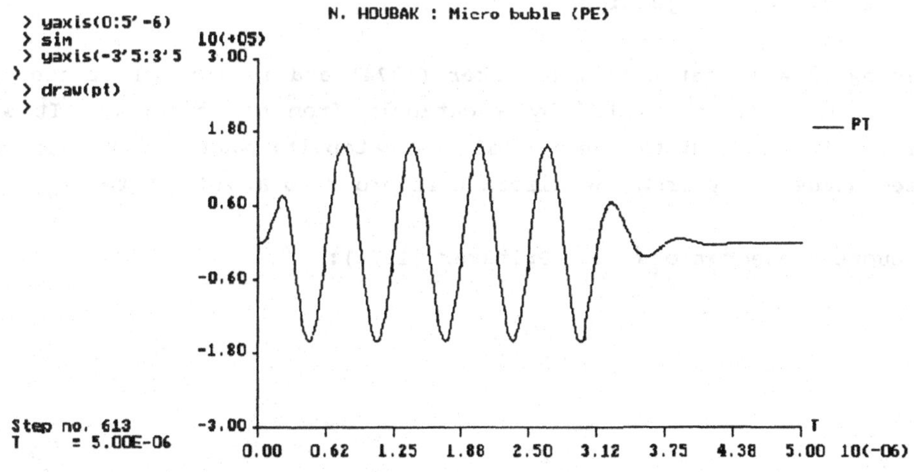

Figure 8.4: The pulse (the forcing function).

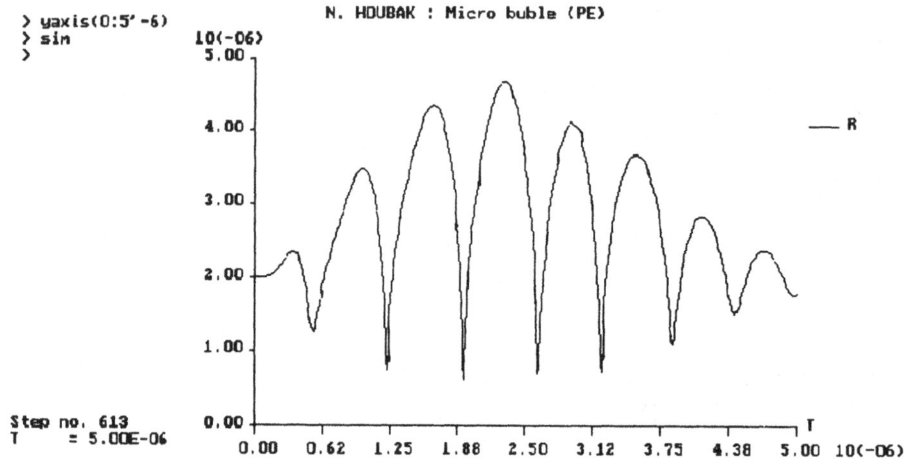

Figure 8.5: Radius of μbubble versus time.

The figures 8.4 and 8.5 show the shape of the pulse (8.4) and the resulting vibration of the bubble (8.5). The results agree very accurately with the line printer plots in Lewin and Bjørnø (1982).

Example 2: Pilot ejection.

This example is taken from Pritsker (1974) and is a model of the behavior of a pilot ejected by a catapult from his aircraft. It was originally used in the design of the catapult; what is the minimum speed needed away from the aircraft in order to avoid the tail?

We quote the equations from Pritsker (1974):

$$\frac{dX}{dt} = V\cos(\theta) - V_A \qquad -60 \le X \le 0$$

$$\frac{dY}{dt} = V\sin(\theta)$$

$$\frac{dV}{dt} = 0 \qquad \text{for} \quad 0 \le Y < Y_1$$

$$\frac{dV}{dt} = -\frac{D}{M} - g\sin(\theta) \qquad \text{for} \quad Y_1 \le Y$$

$$\frac{d\theta}{dt} = 0 \qquad \text{for} \quad 0 \le Y < Y_1$$

$$\frac{d\theta}{dt} = -\frac{g\cos(\theta)}{V} \qquad \text{for} \quad Y_1 \le Y$$

$$D = \frac{1}{2}\rho\, C_d S\, V^2$$

The initial values are as follows

$$V(0) = \sqrt{(V_A - V_E\sin(\theta_E))^2 + (V_E\cos(\theta_E))^2}$$

$$\theta(0) = \tan^{-1}\left(\frac{V_E\cos(\theta_E)}{V_A - V_E\sin(\theta_E)}\right)$$

$$X(0) = 0, \qquad Y(0) = 0$$

and in the simulation we have used the following values for the parameters in the model (English units):

$C_d = 1.0$, $g = 32.2$ ft/sec^2, $\theta_E = 0.2618$ rad
$\rho = 2.3769 \cdot 10^{-3}$ slugs/ft^3, $S = 10$ ft^2, $Y_1 = 4$ ft
$V_A = 900$ ft/sec, $V_E = 40$ ft/sec, $M = 7$ slugs.

This model requires the following facilities in the SIL language:

1) handling of a discontinuity,
2) computing initial values based on parameters,
3) trigonometric functions, and
4) termination of the simulation based on a variable exceeding its range.

```
$TITLE= Pilot ejection model.
begin

(*   Declarations    *)
variable    x_pos(0)(-60:0), y_pos(0), speed(), theta(),
            a, b, d;

parameter   cd(1), g(32.2), rho(2.3769'-3), va(900),
            ve(40), thetrad(0.2618), xm(7), xs(10),
            y1(4);

derivative dx_pos(x_pos),  dy_pos(y_pos),
           speeddot(speed), thetadot(theta);

time t(0:2);

a := va-ve*sin(thetrad);              (*  auxillary  *)
b := ve*cos(thetrad);                 (*  variables  *)

speed := sqrt( a ** 2 + b ** 2 );  (* initial values *)
theta := atan( b / a );            (* are calculated *)
abserror := 0;      relerror := 5.'-6;

(*  Specify the equations  *)
dx_pos := speed*cos(theta) - va;
dy_pos := speed*sin(theta);

if y_pos < y1 then
   begin      (*  moving out of the aircraft  *)
   speeddot := 0;
   thetadot := 0;
   d := speed
   end
else
   begin       (*  free from the aircraft  *)
   d := rho*cd*xs*speed*speed/2;
   speeddot := -(d/xm + g*sin(theta));
   thetadot := -(g*cos(theta))/speed
   end;

write(200,x_pos,y_pos,speed,theta)
end.
```

Figure 8.6: SIL model for the pilot ejection problem.

The SIL model for solving this problem is shown in fig. 8.6; the model includes the setting of the error tolerances to values specified in Pritsker (1974). Also, it demonstrates how to stop the simulation if a variable (x_pos) exceeds its legal range [-60,0] (the tail is located 60 ft behind the cockpit). The results are shown in the figures 8.7 and 8.8.

S I L   S I M U L A T I O N   R E S U L T S

T	X_POS	Y_POS	SPEED	THETA
0.00000E+0000	0.00000E+0000	0.00000E+0000	8.90486E+0002	4.34023E-0002
1.00000E-0002	-1.03528E-0001	3.86370E-0001	8.90486E+0002	4.34023E-0002
2.00000E-0002	-2.07056E-0001	7.72741E-0001	8.90486E+0002	4.34023E-0002
3.00000E-0002	-3.10584E-0001	1.15911E+0000	8.90486E+0002	4.34023E-0002
4.00000E-0002	-4.14111E-0001	1.54548E+0000	8.90486E+0002	4.34023E-0002
5.00000E-0002	-5.17639E-0001	1.93185E+0000	8.90486E+0002	4.34023E-0002
6.00000E-0002	-6.21167E-0001	2.31822E+0000	8.90486E+0002	4.34023E-0002
7.00000E-0002	-7.24695E-0001	2.70459E+0000	8.90486E+0002	4.34023E-0002
8.00000E-0002	-8.28223E-0001	3.09096E+0000	8.90486E+0002	4.34023E-0002
9.00000E-0002	-9.31751E-0001	3.47733E+0000	8.90486E+0002	4.34023E-0002
1.00000E-0001	-1.03528E+0000	3.86370E+0000	8.90486E+0002	4.34023E-0002
1.10000E-0001	-1.16681E+0000	4.24818E+0000	8.81848E+0002	4.31674E-0002
1.20000E-0001	-1.42184E+0000	4.62431E+0000	8.68826E+0002	4.27998E-0002
1.30000E-0001	-1.80492E+0000	4.99174E+0000	8.56183E+0002	4.24268E-0002
1.40000E-0001	-2.31237E+0000	5.35067E+0000	8.43902E+0002	4.20483E-0002
1.50000E-0001	-2.94063E+0000	5.70129E+0000	8.31969E+0002	4.16644E-0002
1.60000E-0001	-3.68632E+0000	6.04380E+0000	8.20368E+0002	4.12749E-0002
1.70000E-0001	-4.54620E+0000	6.37837E+0000	8.09086E+0002	4.08800E-0002
1.80000E-0001	-5.51712E+0000	6.70518E+0000	7.98109E+0002	4.04796E-0002
1.90000E-0001	-6.59611E+0000	7.02439E+0000	7.87426E+0002	4.00738E-0002
2.00000E-0001	-7.78030E+0000	7.33616E+0000	7.77026E+0002	3.96624E-0002
2.10000E-0001	-9.06693E+0000	7.64065E+0000	7.66896E+0002	3.92456E-0002
2.20000E-0001	-1.04533E+0001	7.93800E+0000	7.57027E+0002	3.88233E-0002
2.30000E-0001	-1.19370E+0001	8.22836E+0000	7.47408E+0002	3.83956E-0002
2.40000E-0001	-1.35154E+0001	8.51185E+0000	7.38031E+0002	3.79623E-0002
2.50000E-0001	-1.51862E+0001	8.78861E+0000	7.28885E+0002	3.75236E-0002
2.60000E-0001	-1.69472E+0001	9.05877E+0000	7.19964E+0002	3.70794E-0002
2.70000E-0001	-1.87961E+0001	9.32245E+0000	7.11258E+0002	3.66297E-0002
2.80000E-0001	-2.07309E+0001	9.57976E+0000	7.02760E+0002	3.61746E-0002
2.90000E-0001	-2.27495E+0001	9.83081E+0000	6.94463E+0002	3.57140E-0002
3.00000E-0001	-2.48499E+0001	1.00757E+0001	6.86359E+0002	3.52478E-0002
3.10000E-0001	-2.70302E+0001	1.03146E+0001	6.78442E+0002	3.47763E-0002
3.20000E-0001	-2.92886E+0001	1.05475E+0001	6.70705E+0002	3.42992E-0002
3.30000E-0001	-3.16234E+0001	1.07746E+0001	6.63143E+0002	3.38166E-0002
3.40000E-0001	-3.40328E+0001	1.09960E+0001	6.55749E+0002	3.33286E-0002
3.50000E-0001	-3.65152E+0001	1.12117E+0001	6.48518E+0002	3.28351E-0002
3.60000E-0001	-3.90689E+0001	1.14218E+0001	6.41445E+0002	3.23361E-0002
3.70000E-0001	-4.16925E+0001	1.16265E+0001	6.34525E+0002	3.18316E-0002
3.80000E-0001	-4.43844E+0001	1.18257E+0001	6.27752E+0002	3.13217E-0002
3.90000E-0001	-4.71431E+0001	1.20197E+0001	6.21122E+0002	3.08063E-0002
4.00000E-0001	-4.99673E+0001	1.22084E+0001	6.14631E+0002	3.02853E-0002
4.10000E-0001	-5.28557E+0001	1.23919E+0001	6.08274E+0002	2.97590E-0002
4.20000E-0001	-5.58068E+0001	1.25704E+0001	6.02047E+0002	2.92271E-0002
4.30000E-0001	-5.88195E+0001	1.27438E+0001	5.95946E+0002	2.86897E-0002
4.40000E-0001	-6.18924E+0001	1.29123E+0001	5.89968E+0002	2.81469E-0002

9.23 SECONDS in execution

159.5 KBytes left in Long Heap memory

Figure 8.7: Tabular solution to the pilot ejection model; the tabel is comparable to fig. 6-11 in Pritsker (1974).

Fig. 8.7 contain the solution as a table which is comparable with a similar table in Pritsker (1974); the solution generated by SIL has at the endpoint 2 digits in common with the solution found by Pritsker (1974). The plot generated in fig. 8.8 demonstrates how to reverse the X-axis, the error message displayed when HELP is used, and how to put options on the axes. The discontinuity (at Y_pos = 4 ft) can be located on the graph. The time used for solving the problem is about 10 seconds.

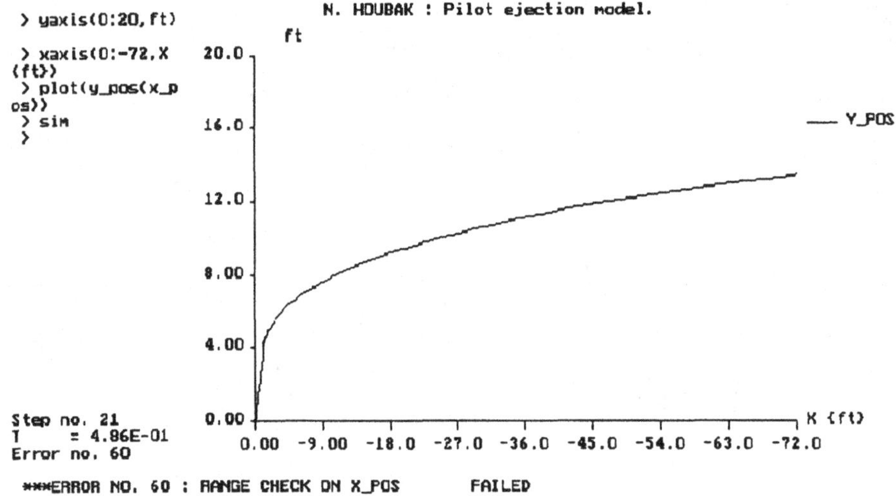

Figure 8.8: Graphic screen dump from running the pilot ejection model.

Example 3: Asynchronous electric motor.

This problem is part of a larger exercise originating from a course lectured by O. Jensen at the Danish Engineering Academy on the dynamics of electric mashinery. Further information can be obtained from Drønen (1985).

The time dependent 3-phase voltages can be computed as

$$U_a = U \cdot \sin(2\pi f \cdot t)$$
$$U_b = U \cdot \sin(2\pi f \cdot t - 2\pi/3)$$
$$U_c = U \cdot \sin(2\pi f \cdot t + 2\pi/3)$$

where $f = 50$ Hz is the frequency, t is the running time, and U = 310.27 is the peak value of the 220 Volts power supply. These voltages can be transformed to a Q-D plane by the formulae

$$U_d = (2*U_a - U_b - U_c)/3$$
$$U_q = -\sqrt{3}\,(U_b - U_c)/3 \ .$$

For the stator we have the following two implicitly given ordinary differential equations,

$$0 = r_1\, i_{d1} + L_{11}\frac{d\, i_{d1}}{dt} + L_{12}\frac{d\, i_{d2}}{dt} - U_d$$

$$0 = r_1\, i_{q1} + L_{11}\frac{d\, i_{q1}}{dt} + L_{12}\frac{d\, i_{q2}}{dt} - U_q$$

with R_1, L_{11}, and L_{12} being a resistance and inductances, respectively. i_{d1} and i_{q1} are two currents. We have two almost similar equations for the rotor part of the motor,

$$0 = r_2\, i_{d2} + L_{22}\frac{d\, i_{d2}}{dt} + L_{12}\frac{d\, i_{d1}}{dt} - \omega\,(i_{q2}L_{22}+i_{q1}L_{12})$$

$$0 = r_2\, i_{q2} + L_{22}\frac{d\, i_{q2}}{dt} + L_{12}\frac{d\, i_{q1}}{dt} + \omega\,(i_{d2}L_{22}+i_{d1}L_{12})$$

where ω is the angular velocity of the rotor. The air gap momentum M can be determined by the following equation

$$M = 3\, L_{12}\,(i_{d1}i_{q2} - i_{d2}i_{q1})$$

and with this, the following equation determines the (mechanical) motion of the rotor,

$$\omega' = 2(M - T_L)/(I_M + I_L) \ .$$

Here T_L is the (time dependent) load of the motor, and I_M and I_L are the moments of inertia for the motor and the load, respectively.

Putting all these equations together and adding the values of all the parameters one obtain the following SIL model of the system. The model is shown in fig. 8.9.

```
$TITLE= Dynamics of async. motor
BEGIN

PARAMETER R1(0.4715), R2(0.6945),
          L1(4.56E-3), L2(4.56E-3),
          L12(96.49E-3), L11(0.101), L22(0.101),
          PI(3.14159),
          F(50), U(310.27), IM(0.0267), IB(0.061);

VARIABLE  TB, UA, UB, UC, UD1, UQ1, M, TWOPIF, PI23,
          ID1(0), IQ1(0), ID2(0), IQ2(0), OM(0);

DERIVATIVE DID1(ID1)(0), DIQ1(IQ1)(0),
           DID2(ID2)(0), DIQ2(IQ2)(0),
           DOM(OM);

TIME T(0:0.8);
METHOD := 139;  ABSERROR := 1.0E-5;  RELERROR := 1.0E-3;

(*         Time dependent load    *)
IF T < 0.3 THEN
   TB := 0
ELSE
   IF T < 0.7 THEN
      TB := 1200 * (T - 0.3) / 0.4
   ELSE
      TB := 1200;

(*         Stator voltage          *)
TWOPIF := 2*PI*F;       PI23   := 2*PI/3;
UA := U*SIN(TWOPIF*T);
UB := U*SIN(TWOPIF*T - PI23);
UC := U*SIN(TWOPIF*T + PI23);

(*         Voltage transformations
           from 3 phase to axis values  *)
UD1 := (2*UA - UB - UC)/3;
UQ1 := -1.7321*(UB - UC)/3;

(*         Stator equations (ODEs)  *)
1.0E-8 := R1*ID1 + L11*DID1 + L12*DID2 - UD1;
1.0E-8 := R1*IQ1 + L11*DIQ1 + L12*DIQ2 - UQ1;

(*         Rotor equations  (ODEs)  *)
1.0E-8 := R2*ID2 + L22*DID2 + L12*DID1 -
                       OM*(IQ2*L22 + IQ1*L12);
1.0E-8 := R2*IQ2 + L22*DIQ2 + L12*DIQ1 +
                       OM*(ID2*L22 + ID1*L12);

(*         Air gap momentum         *)
M := 3*L12*(ID1*IQ2 - ID2*IQ1);
(*         Rotor angle equation     *)
DOM := 2*(M - TB)/(IM + IB);

(*         Output specification     *)
WRITE(300,TB,ID1);
WRITE(300,M(OM))
END.
```

Figure 8.9: SIL model of an asynchronous electrical motor.

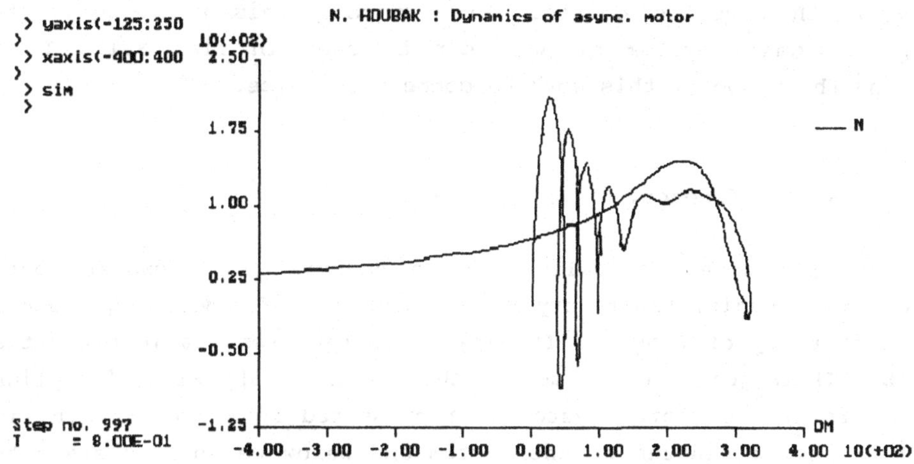

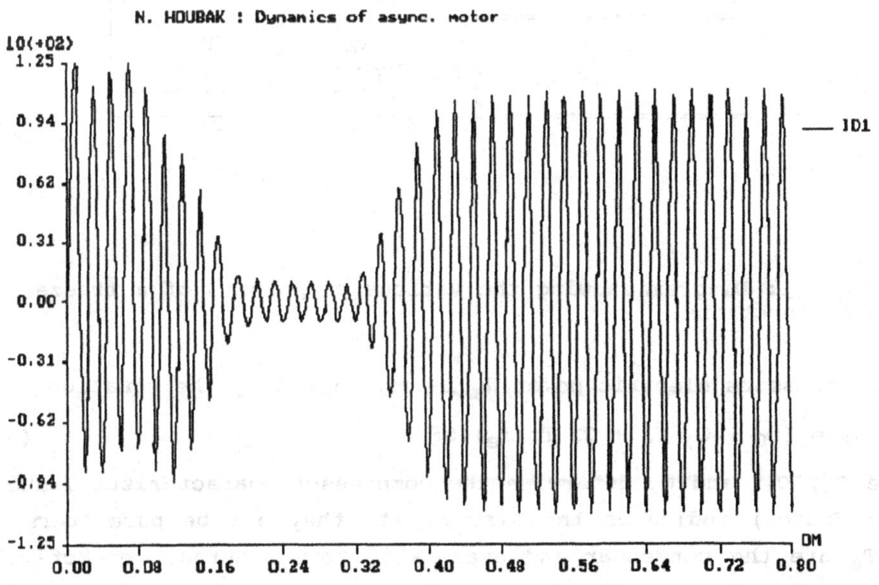

<u>Figure 8.10</u>: Screen dumps obtained from running the asynchronuos electrical engine model.

Some of the results obtained from running this model using the graphic command system of μSIL can be seen in fig. 8.10. It is beyound the scope of this work to comment on these.

Example 4: The freezer.

This example comes basically from an exercise in a master course given by P. Danig, Laboratory for Cooling and refrigerating, Technical University of Denmark. It will illustrate some basic properties of the SIL language to be used in the design of any kind of appliances; here the simulated system is represented by a freezer. In fig. 8.11 we show a diagram of the system and below we shortly state the equations for its various components.

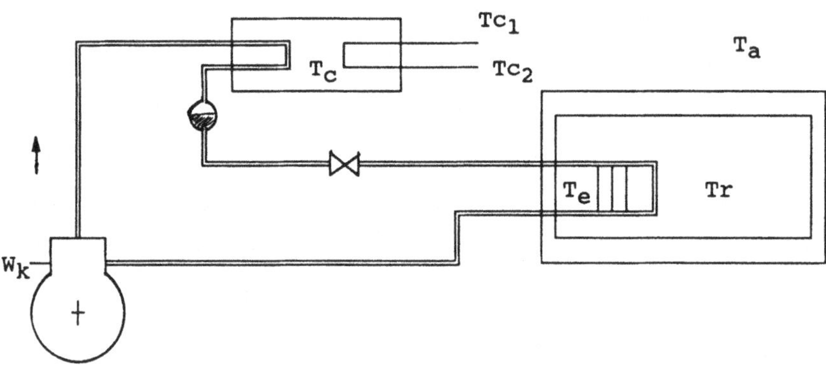

Figure 8.11: Diagram showing the various components of a freezer and their connenctions.

The compressors electric power W_k is determined by the equation

$$W_k = (C_0 + C_1\ T_c + C2\ Tc\ T_e)\ SF \qquad (8.1)$$

where C_0, C_1, and C_2 determine the compressor characteristic, and SF (Size Factor) indicates the size of it; they all be parameters. T_c and T_e are the condenser and evaporator temperatures, respectively. The condensor heat Q_c is determined by

$$Q_c = W_k\ Carn\ (T_c + 273.15)/(T_c - T_e) \qquad (8.2)$$

where Carn is the Carnot efficiency - a parameter. The heat can also be determined by

$$Q_c = Flc\ Cp_w\ (T_{c2} - T_{c1}) \tag{8.3}$$

where Flc and Cp_w are parameters. The temperature increase of the cooleant is given by

$$(T_{c2} - T_{c1}) = (T_c - T_{c1})\ Ec. \tag{8.4}$$

At the evaporator the heat Q_e transferred from the air in the freezer to the cooleant can be expressed as

$$Q_e = U_a\ (T_r - T_e) \tag{8.5}$$

where T_r is the room temperature and U_a is a heat transfer coefficient. The heat balance in the room can be expressed as follows,

$$Q_e = Aka\ (T_a - T_r) + Q_b \tag{8.6}$$

where T_a is the ambient temperature, Aka is a heat transfer coefficient, and Q_b is an internal heat production. The last equation describes the overall energy balance in the system

$$W_k + Q_e = Q_c. \tag{8.7}$$

In the above 7 equations ((8.1) to (8.7)) many of the quantities are parameters, below we give a list with their values:

Aka	=	0.18	kW/°C	C_0	=	0.5895	kW
C_1	=	0.010264	kW/°C	C_2	=	$2.869\ 10^{-4}$	kW/°C^2
Carn	=	0.45		Cp_w	=	4.2	kJ/kg/°C
E_c	=	0.7		Flc	=	0.40334	kg/s
Q_b	=	1.5	kW	SF	=	13.0	
T_a	=	25.0	°C	Tc_1	=	20.0	°C
U_a	=	1.7	kW/°C				

The remaining 7 variables (W_k, Q_c, Tc_2, T_c, T_e, T_r, and Q_e) are then to be determined on basis of the 7 equations; we call them static variables. Due to the nature of the equations it is though impossible to evaluate any of the variables based only on the parameters; the equations contain an algebraic loop and the variables are implicitly given static variables.

A SIL program which implements the model can be seen in fig. 8.12. Notice in the model the following details,

1) all 7 equations are given in residual form: 0 = f(...). This is indicated by giving the residual error bound as the left hand side of the 'assignment statement'.
2) the order of the equations has been changed but this will not influence the solution.
3) equation 8.2 has been rearranged in order to avoid a division with a temperature difference; it is good practice to avoid divisions because they are computationally expensive and they can introduce a division by zero error during the solution process.
4) all the declared implicit variables have an initial value associated; these values are the initial guess used in the Newton process.

```
BEGIN    (*  Freezer model; static simulation    *)

(*  Declarations      *)
PARAMETER    AKA(0.18),                  C0(0.5895),
             C1(0.010264),               C2(2.869'-4),
             CARN(0.45),                 CPW(4.2),
             EC(0.7),                    FLC(0.40334),
             QB(1.5),                    SF(13),
             TA(25),     TC1(20),        UA(1.7);

(*   Define static IMPLICIT variables  *)
VARIABLE     TC(25),                     TC2(25.1),
             TE(-40),                    TR(-20),
             QC(18),     QE(10),         WK(8);

(*   Define the equations in residual form    *)
1.0'-5  := QC -(QE + WK);
1.0'-5  := WK*CARN*(TC+273.15) - QC*(TC-TE);
1.0'-5  := FLC*CPW*(TC2-TC1) - QC;
1.0'-5  := (TC - TC1)*EC - (TC2 - TC1);
1.0'-5  := UA*(TR - TE) - QE;
1.0'-5  := SF*(C0 + C1*TC + C2*TC*TE) - WK;
1.0'-5  := QB + AKA*(TA-TR) - QE;

(*   Output specifications     *)
WRITE(TR,WK,QE)
END.
```

Figure 8.12: SIL model of a freezer/refrigerator describing the static behavior.

The results generated in the .LST file after less than 10 seconds of computing time are as in fig. 8.13. These results says that the stationary temperature in the freezer will be -22.5 °C when the compressor is running all the time.

```
MODEL CONSISTS OF :
  13 PARAMETERS
   7 IMPLICIT STATIC VARIABLES

SIMULATION STATISTICS:

   NUMBER OF ACCEPTED STEPS        :    0
   TOTAL NUMBER OF FUNCTION CALLS  :    1
   NUMBER OF ALGEBRAIC ITERATIONS  :   13

SIMULATION OPTIONS USED:

   DEBUG               :    1
   METHOD              :  119
   MAXORDER            :   10
   MAXCPU              :  0.0
   INITIAL TIME        : 0.000E+0000
   FINAL TIME          : 0.000E+0000
   MAXIMUM STEPSIZE    : 0.000E+0000
   INITIAL STEPSIZE    : 0.000E+0000
   ABSERROR            : 1.000E-0005
   RELERROR            : 1.000E-0005

VALUES OBTAINED DURING SIMULATION:

   MAXORDER            :    0
   MAXIMUM STEPSIZE    : 0.000E+0000
   MINIMUM STEPSIZE    : 0.000E+0000


      S I L   S I M U L A T I O N   R E S U L T S
     Time            TR              WK              QE
0.00000E+0000 -2.25462E+0001  8.63903E+0000  1.00583E+0001

  6.81  SECONDS in execution

 160.0  KBytes left in Long Heap memory
```

Figure 8.13: The .LST file obtained from running the SIL model in fig. 8.12.

The parameter SF determines the size of the compressor; let us investigate how the statinary temperature in the freezer varies when we change the size of the compressor. Though it is impossible to change the size of the compressor continuously we will vary SF in the interval [4,20]. We do that by declaring SF as a TIME variable varying between 4 and 20 instead of being a parameter. The resulting SIL model is in fig. 8.14.

```
BEGIN
(* Refrigerator model; quasi-static simulation *)
PARAMETER    AKA(0.18),              C0(0.5895),
             C1(0.010264),           C2(2.869'-4),
             CARN(0.45),             CPW(4.2),
             EC(0.7),                FLC(0.40334),
             QB(1.5),    (* delete   SF(13),    *)
             TA(25),                 TC1(20),
             UA(1.7);

(*   Define static IMPLICIT variables   *)
VARIABLE     TC(25),                 TC2(25.1),
             TE(-40),                TR(-20),
             QC(18),                 QE(10),
             WK(8);

(*   Define the 'TIME' variable   SF    *)
TIME SF(4:20);

(*   Define the equations in residual form     *)
1.0'-5 := QC -(QE + WK);
1.0'-5 := WK*CARN*(TC+273.15) - QC*(TC-TE);
1.0'-5 := FLC*CPW*(TC2-TC1) - QC;
1.0'-5 := (TC - TC1)*EC - (TC2 - TC1);
1.0'-5 := UA*(TR - TE) - QE;
1.0'-5 := SF*(C0 + C1*TC + C2*TC*TE) - WK;
1.0'-5 := QB + AKA*(TA-TR) - QE;

(*   Output specifications     *)
WRITE(TR)
END.
```

Figure 8.14: SIL model for Quasi-static simulation of a freezer.

When running this model with μSIL being in the graphic mode it takes about 30 seconds to obtain the curve shown in fig 8.15. It is also possible in the graphic mode to change the value of for example Aka (the insulation) and then see how this influences the behavior of the stationary solution.

Finally, SF is set to 13 and we want to analyze the dynamic behavior of the system. Assume the freezer contains a certain mass of meet with given heat capacity (Mc = 20000.0 kJ/°C) and that its temperature equals the room temperature T_r. Equation 8.6 can then be extendend to take into account the change of temperature in the freezer. This gives us

$$T_r' = (Aka\ (T_a - T_r) + Q_b - Q_e)/M_c$$

where T_r' is the time derivative of the room temperature T_r. Now T_r become a dynamic variable. A suitable range for the simulation time is $[0, 10^5]$ (seconds).

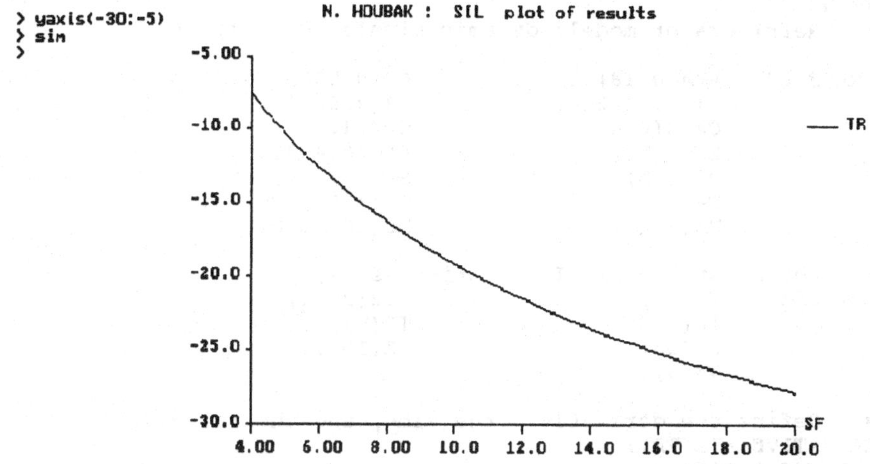

Figure 8.15: Graphic screen dump from running the Quasi stationary SIL model of a freezer (temperature versus size of compressor).

We also assume that the freezer is equipped with an ON-OFF temperature control operating in the range [-20,-18]. It turns ON the compressor when T_r comes above -18 and turns it OFF when T_r comes below -20. This implemented in the SIL model using the IF THEN ELSE construction with a SWITCH variable simulating the relay. The total model in the SIL language is shown in fig 8.16.

After running this model about 3 minutes in the graphic command mode of the μSIL system the screen looks as in fig. 8.17. The behaviour of the ON-OFF control is very clear from this graph.

```
BEGIN
(*   Refrigerator model; dynamic simulation   *)

PARAMETER    AKA(0.18),                  C0(0.5895),
             C1(0.010264),               C2(2.869'-4),
             CARN(0.45),                 CPW(4.2),
             EC(0.7),                    FLC(0.40334),
             QB(1.5),                    SF(13),
             TA(25),                     TC1(20),
             UA(1.7),                    MC(20000.0);

(*   Define static IMPLICIT variables   *)
VARIABLE     TC(25),                     TC2(25.1),
             TE(-40),                    TR(-15),
             QC(18),                     QE(10),
             WK(8);

(*   Define the derivative, the time, and the relay *)
DERIVATIVE DTR(TR);
TIME T(0:1.0'5);
SWITCH RELAY(ON);

(*   Define the equations in residual form   *)
1.0'-5 := QC -(QE + WK);
1.0'-5 := WK*CARN*(TC+273.15) - QC*(TC-TE);
1.0'-5 := FLC*CPW*(TC2-TC1) - QC;
1.0'-5 := (TC - TC1)*EC - (TC2 - TC1);
1.0'-5 := UA*(TR - TE) - QE;
IF RELAY THEN
   BEGIN    (*  compressor turned  ON    *)
   RELAY := TR > -20;
   1.0'-5 := SF*(C0 + C1*TC + C2*TC*TE) - WK
   END
ELSE
   BEGIN    (*  compressor turned OFF    *)
   RELAY := TR > -18;
   1.0'-5 := WK
   END;

(*   Define the differential equation   *)
DTR := (QB + AKA*(TA-TR) - QE)/MC;

(*   Output specifications    *)
PRINT(TR,WK,QE,QC)
END.
```

Figure 8.16: Dynamic SIL model of a freezer with ON-OFF temperature control.

With the concluding remark that taking into account the temperature and heat capacity of the air in the freezer will cause the system of ordinary differential equations to become stiff, this is the end of example 4.

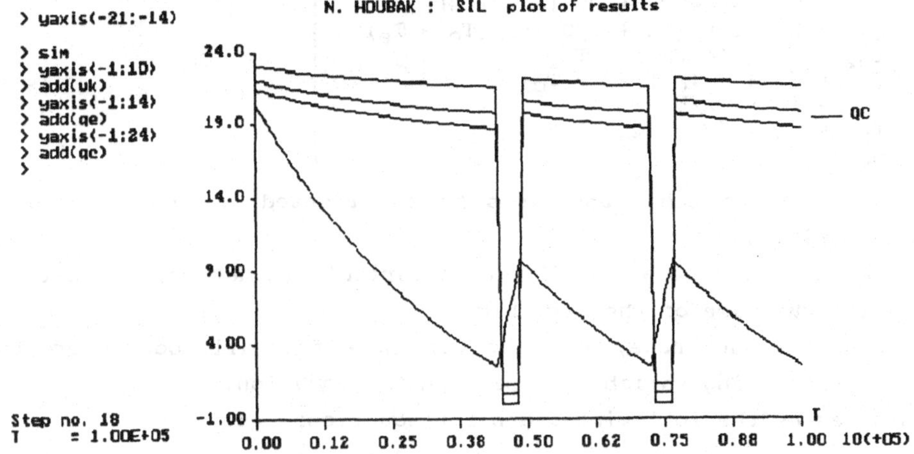

Figure 8.17: Dump of the graphic screen after running the dynamic freezer model.

Example 5: Model analysis.

In this example we will analyze the model from the previous example in order to reduce the number of implicit equations. The technique can be used gain insight in the couplings in any system of coupled equations.

The equations 8.1 to 8.7 are rewritten in the form

<id> := <expression>

(assignment statements) with the restriction that each variable is assigned once. This gives us equations 8.8.

$$\left.\begin{aligned}
W_k &= (C_0 + C_1\ T_c + C2\ Tc\ T_e)\ SF \\
Q_c &= W_k\ Carn\ (T_c + 273.15)/(T_c - T_e) \\
T_{c2} &= Q_c/(Flc\ Cp_w) + T_{c1} \\
T_c &= (T_{c2} - T_{c1})/Ec + T_{c1} \\
T_e &= T_r - Q_e/U_a \\
T_r &= T_a + (Q_b - Q_e)/Aka \\
Q_e &= Q_c - W_k
\end{aligned}\right\} \qquad (8.8)$$

We can represent these equations by an oriented graph generated in the following way:

1) There should be a node for each variable (equation); mark these with the name of the variable.
2) Draw for each node, edges to that node from all nodes where the corresponding variables appear in the equation.

This gives us the following graph for equations 8.8.

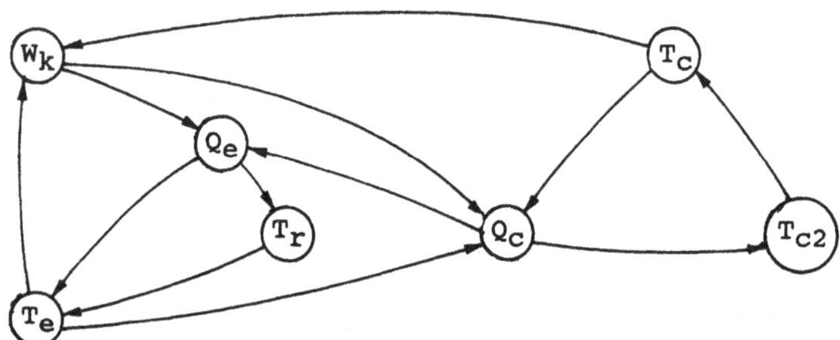

Figure 8.18: Graph of the connections in equations 8.8.

The graph contains 10 oriented loops but only 2 simultaneously independent loops. Two loops are independent if they have no common node. The 2 loops contain the nodes Q_c, T_{c2}, and T_c and W_k, Q_e, (T_r,) and T_e. One node from each of the loops must be an implicit variable but the remaining can be declared and assigned as special explicit variables, which depends on implicit variables (see chapter 6.2.3 Implicit assignment).

In this example, T_r cannot be used as an implicit variable since it is not a common node to all loops in one of the two sets of disjoint loops. We have chosen Q_c and T_e to be our implicit variables. This gives the SIL model shown in fig. 8.19.

The solution to this model is exactly the same as the one shown in fig. 8.13. The only difference between the two models is the computing time; the solution to this model is found after only 2 seconds of computing time and that is a factor of three less than the time used for solving the original problem. For this problem, most of the computing time is spent on building, factorizing, and solving the iteration matrix and this cost is reduced significantly when the order of the matrix is reduced from 7 to 2.

```
BEGIN
(*   Refrigerator model; static simulation    *)

PARAMETER    AKA(0.18),                 C0(0.5895),
             C1(0.010264),              C2(2.869'-4),
             CARN(0.45),                CPW(4.2),
             EC(0.7),                   FLC(0.40334),
             QB(1.5),                   SF(13),
             TA(25),                    TC1(20),
             UA(1.7);

(*   Define static IMPLICIT variables   *)
VARIABLE     TC(  ),                    TC2(  ),
             TE(-40),                   TR(  ),
             QC(18),                    QE(  ),
             WK(  );

(*   Define the explicit equations    *)
TC2  := TC1 + QC/(FLC*CPW);
TC   := TC1 + (TC2 - TC1)/EC;
WK   := SF*(C0 + C1*TC + C2*TC*TE);
QE   := QC - WK;
TR   := (QB - QE)/AKA + TA;

(*   Define the implicit equations in residual form    *)
1.0'-5 := UA*(TR - TE) - QE;
1.0'-5 := WK*CARN*(TC+273.15) - QC*(TC-TE);

(*   Output specifications     *)
WRITE(TR,WK,QE,QC)
END.
```

Figure 8.19: A SIL model with a reduced number of implicit equations.

Example 6: A chemical model (combustion).

This model comes from a PhD project carried out at the Laboratory for Energetics by Nielsen (1986). The project concerned the problem of describing theoretically and practically the observed effect of adding Hydrogen to the intake air of an diesel engine; the engine tends to ignition failures when small amounts of Hydrogen are added.

There exists several models of the combustion process of both hydrocarbon fuels and Hydrogen. A major task in the theoretical part of this project was to combine these models (one for hydrocarbon and one for Hydrogene) and then study the interference between the models. It was essential that the combined model predicts the results obtained in the practical part of the project. This example is the most simple of the Hydrogene combustion models, the interested reader is referred to Nielsen (1986) for further details.

The Hydrogene combustion model is shown in fig. 8.20, and the equations there are mainly specifications of reaction rates for the different species and the reaction equations for the process. The problem is a purely dynamic problem; all equations can in principles be specified as ordinary differential equations. The model is a stiff system of ODE's and it therefore requires the built-in variable METHOD to be assigned the value 139. One can observe the effect of the stiffness on the stepsizes used; they vary from 10^{-13} to 10^{-3} (10 orders of magnitude).

```
BEGIN
(* Cumbustion model based on Dryer.
   Reverse rate values by Gardiner et al., unless
   otherwise noticed.

   Reactions  (Even numbers refer to reverse rates)
   01  :  H + O2        = OH + O
   03  :  H2 + O        = OH + H
   05  :  O + H2O       = 2*OH
   07  :  H + H2O       = OH + H2
   09  :  O + H + M     = OH + M
   11  :  2*O + M       = O2 + M
   13  :  H2 + M        = 2*H + M
   15  :  H2O + M       = H + OH + M
   17  :  H + O2 + M    = HO2 + M
   19  :  HO2 + O       = O2 + OH
   21  :  H + HO2       = 2*OH
   23  :  H + HO2       = H2 + O2
   25  :  OH + HO2      = H2O + O2
   27  :  2*HO2         = H2O2 + O2
   29  :  H2O2 + M      = 2*OH + M
   31  :  H + H2O2      = HO2 + H2
   33  :  H2O2 + OH     = H2O + HO2            *)

PARAMETER  (*  MOLE FRACTIONS  *)
           FH2(0.296), FN2(0.556), FO2(0.148),
           CR(25.0)   (* COMPRESSION RATIO  *) ;

VARIABLE   (*  REACTION RATES   *)
      R01, R02, R03, R04, R05, R06, R07, R08, R09, R10,
      R11, R12, R13, R14, R15, R16, R17, R18, R19, R20,
      R21, R22, R23, R24, R25, R26, R27, R28, R29, R30,
      R31, R32, R33, R34,
      (*  ABSOLUTE TEMPERATURE,
                    PRESSURE, AND TOTAL CONCENTR. *)
      T, P, TC,
      (*  CONCENTRATIONS IN MOL/CM3  (INITIAL VALUE)   *)
```

```
      H(0.0),
      HO2(0.0),
      H2(),    (*  INITIAL VALUE WILL BE CALCULATED  *)
      H2O(0.0),
      H2O2(0.0),
      N2(),    (*  INITIAL VALUE WILL BE CALCULATED  *)
      O(0.0),
      OH(0.0),
      O2();    (*  INITIAL VALUE WILL BE CALCULATED  *)

DERIVATIVE    (*  OF CONCENTRATIONS  *)
              DH(H),
              DHO2(HO2),
              DH2(H2),
              DH2O(H2O),
              DH2O2(H2O2),
              DO1(O),      DOH(OH),
              DO2(O2),     DN2(N2);

(*  INTEGRATION TIME AND OPTIONS      *)
TIME TID(0:0.01);
METHOD := 139;  ABSERROR := 1E-30;

(*  ABSOLUTE TEMPERATURE IN KELVIN    *)
T  :=  305*EXP(0.36*LOG(CR));
(*  ABSOLUTE PRESSURE    IN BAR       *)
P  := 1.E5*EXP(1.36*LOG(CR));
(*  TOTAL CONCENTRATION  IN MOL/CM3   *)
TC := 1.E-3*P/(8317*T);
(*  INITIAL CONCENTRATIONS IN MOL/CM3*)
H2 := FH2*TC;
N2 := FN2*TC;
O2 := FO2*TC;

(*  REACTION RATES       IN MOL/CM3/S  *)
R01  :=   2.200E14 *EXP( -8.45E3/T)*H*O2;
R02  :=   1.800E13                 *OH*O;
R03  :=   1.82E10*T*EXP( -4.48E3/T)*H2*O;
R04  :=   1.400E12 *EXP( -3.02E3/T)*OH*H;  (* KAUFMAN *)
R05  :=   6.760E13 *EXP( -9.24E3/T)*O*H2O;
R06  :=   1.500E09 *EXP(1.14*LOG(T))*OH**2;
R07  :=   9.330E13 *EXP(-10.55E3/T)*H*H2O;
R08  :=   1.000E08*EXP(1.6*LOG(T))
                  *EXP(-1.66E3/T)*OH*H2;
R09  :=   1.000E16                 *O*H*N2;
R10  :=   0;
R11  :=   5.000E15 *EXP(-0.25*LOG(T))*O**2*N2;
R12  :=   1.200E14 *EXP(-54.20E3/T)*O2*N2;
R13  :=   2.200E14 *EXP(-48.30E3/T)*H2*N2;
R14  :=   6.400E17 /T              *H**2*N2;
R15  :=   2.200E16 *EXP(-52.90E3/T)*H2O*N2;
R16  :=   1.400E23 /T**2           *H*OH*N2;
R17  :=   1.6600E15 *EXP(+0.50E3/T)*H*O2*N2;
R18  :=   0;
R19  :=   5.0000E13 *EXP(-0.50E5/T)*HO2*O;
R20  :=   0;
R21  :=   2.5000E14 *EXP(-0.96E3/T)*H*HO2;
R22  :=   0;
R23  :=   2.5000E13 *EXP(-0.35E3/T)*H*HO2;
R24  :=   0;
R25  :=   5.0000E13 *EXP(-0.05E3/T)*OH*HO2;
```

```
R26   :=   0;
R27   :=   1.0000E13 *EXP(-0.50E3/T)*HO2**2;
R28   :=   0;
R29   :=   6.3000E17 *EXP(-22.9E3/T)*H2O2*N2;
R30   :=   1.3000E22 /T**2           *OH**2*N2;
R31   :=   1.7000E12 *EXP(-1.90E3/T)*H*H2O2;
R32   :=   7.3000E11 *EXP(-9.40E3/T)*HO2*H2;
R33   :=   1.0000E13 *EXP(-0.90E3/T)*H2O2*OH;
R34   :=   0;

(*  THE EQUATIONS   *)
DH     := -R01 + R02 + R03 - R04 - R07 + R08 - R09 + R10
          +2*(R13 - R14)    + R15 - R16 - R17 + R18
          -R21 + R22 - R23 + R24
          -R31 + R32;
DHO2   :=  R17 - R18 - R19 + R20
          -R21 + R22 - R23 + R24 - R25 + R26
          +2*(-R27 + R28)
          +R31 - R32 + R33 - R34;
DH2    := -R03 + R04 + R07 - R08
          -R13 + R14
          +R23 - R24
          +R31 - R32;
DH2O   := -R05 + R06 - R07 + R08
          -R15 + R16
          +R25 - R26
          +R33 - R34;
DH2O2  :=  R27 - R28 - R29 + R30
          -R31 + R32 - R33 + R34;
DO1    :=  R01 - R02 - R03 + R04 - R05 + R06 - R09 + R10
          -2*(R11 - R12)    - R19 + R20;
DOH    :=  R01 - R02 - R03 + R04 + 2*(R05 - R06)
          +R07 - R08 + R09 - R10
          +R15 - R16 + R19 - R20
          +2*(R21 - R22) -   R25 + R26 + 2*(R29 - R30)
          -R33 + R34;
DO2    := -R01 + R02
          +R11 - R12 - R17 + R18 + R19 - R20
          +R23 - R24 + R25 - R26 + R27 - R28;
DN2    := 0;

(*  OUTPUT   *)
WRITE(P,RO2,H2,O);
PRINT(H,HO2,H2,O);
WRITE(H2O,H2O2,OH,O2);
PRINT(H2O,H2O2,OH,O2);
END.
```

Figure 8.20: SIL model of Hydrogene combustion.

In fig. 8.21 there is a dump from the graphical screen after running this model. The computing time is approximately 15 minutes. The figure also shows that the value of the compression ratio has been changed and the model run again. The curves indicates how sensitive the model is to changes in this parameter.

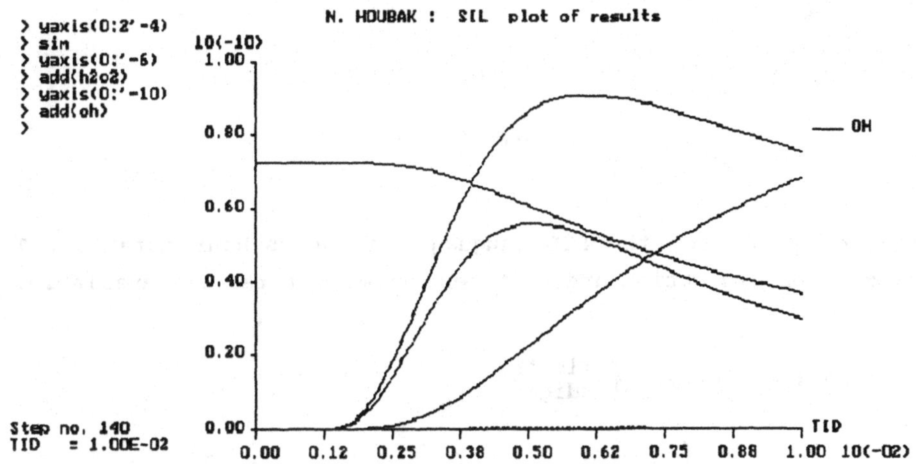

<u>Figure 8.21</u>: Screen dump of the graphics produced by running the SIL model in figure 8.20.

Appendix A.

Complete syntax for the SIL language in Backus-Naur notation. This list may contain option and statement types not readily available.

$$\text{<id> ::= <letter>} \left\{ \begin{array}{l} \text{<letter>} \\ \text{<digit>} \end{array} \right\}_0^*$$

$$\text{<number> ::=} \left\{ \begin{array}{l} \text{<real number>} \left\{ \text{'} \left\{ \begin{array}{c} + \\ - \end{array} \right\}_0^1 \{\text{<int>}\}_0^1 \right\}_0^1 \\ \\ \text{'} \left\{ \begin{array}{c} + \\ - \end{array} \right\}_0^1 \{\text{<int>}\}_0^1 \end{array} \right\}$$

$$\text{<real number> ::=} \left\{ \begin{array}{l} \text{<int>} \{.\ \{\text{<int>}\}_0^1\ \}_0^1 \\ \{\text{<int>}\}_0^1\ .\ \{\text{<int>}\}_0^1 \end{array} \right\}$$

$$\text{<int> ::= <digit>} \{\ \text{<digit>}\ \}_0^*$$

$$\text{<program> ::= <statement> .}$$

$$\text{<statement> ::=} \left\{ \begin{array}{l} \text{<block>} \\ \text{<declaration>} \\ \text{<assignment>} \\ \text{<switch assignment>} \\ \text{<ifthenelse>} \\ \text{<macro call>} \\ \text{<output statement>} \end{array} \right\}$$

$$\text{<block> ::= BEGIN <statement list> END}$$

$$\text{<statement list> ::= <statement>} \{\ ;\ \text{<statement>}\}_0^*$$

$$\text{<declaration> ::=} \left\{ \begin{array}{l} \text{<sampletime declaration>} \\ \text{<variable declaration>} \\ \text{<derivative declaration>} \\ \text{<parameter declaration>} \\ \text{<time declaration>} \\ \text{<discrete declaration>} \\ \text{<switch declaration>} \\ \text{<macro declaration>} \end{array} \right\}$$

```
<parameter declaration> ::= PARAMETER <par id> {, <par id>}^*_0

<par id> ::= <id> ( <initval> )

<initval> ::= {+ | -}^1_0 <number>

<time declaration> ::= TIME <id> { ( <range> ) }^1_0

<range> ::= { {+ | -}^1_0 <number> : { {+ | -}^1_0 <number> }^1_0
            |                      : {+ | -}^1_0 <number> }

<variable declaration> ::=  VARIABLE <var id> {, <var id>}^*_0

<var id> ::= <id> { ( { <range> | <initval> | < <tol> }^1_0 ) }^3_0

<tol> ::= <number>

<derivative declaration> ::= DERIVATIVE <der id> {, <der id>}^*_0

<der id> ::= <id> ( <der of> ) { ( { <initval> | <range> }^1_0 ) }^2_0

<der of> ::= <id>

<switch declaration> ::= SWITCH <switch id> {, <switch id>}^*_0

<switch id> ::= <id> { ( <switch initval> ) }^1_0

<switch initval> ::= { ON | OFF }

<macro declaration> ::= MACRO <macro heading> ;
                              <statement>

<macro heading> ::= <id> { ( <formal parm list> ) }^1_0
```

```
                                                                  *
<formal parm list> ::=  <formal parm> {; <formal parm>}
                                                                  0

                         ┌ ┌ VARIABLE   ┐              * ┐
<formal parm> ::=        │ │ DERIVATIVE │ <id> {, <id>}  │
                         │ └ PARAMETER  ┘              0 │
                         └ TIME <id>                     ┘

                                                                           *
<sampletime declaration> ::= SAMPLETIME <samp id> {, <samp id>}
                                                                           0

<samp id> ::= <id> ( <number> )

                                                                      *
<discrete declaration> ::= DISCRETE <discr id> {, <discr id>}
                                                                      0

<discr id> ::= <id> ( {<number>} ) ( <id> )
                      {<id>    }

<assignment> ::= <lefthand side> :=  <expression>

<lefthand side> ::= { <id>     }
                    { <number> }

                      1                      *
<expression> ::= {+}    <term> { {+} <term> }
                 {-}           { {-}        }
                      0                      0

                                    *
<term> ::= <factor> { {*} <factor> }
                    { {/}          }
                                    0

                                1
<factor> ::= <primary> {** <int>}
                                0

               ┌ <noise generator>           ┐
               │ <number>                    │
<primary> ::=  │ <id>                        │
               │ <std function designator>   │
               └ ( <expression> )            ┘

                                          1            1  1
<noise generator> ::= NOISE { ( {<seed>}   {, <type>} ) }
                                          0            0  0

<seed> ::= <int>

<type> ::= <int>

<std function designator> ::= <std function id> ( <expression> )

                         ┌ SIN  ┐
                         │ COS  │
<std function id> ::=    │ ATAN │
                         │ EXP  │
                         │ LOG  │
                         └ SQRT ┘
```

```
<ifthenelse> ::= IF <switch expr> THEN <statement> ELSE
                                    <statement>

                                                      *
<switch expr> ::= <switch term> { OR <switch term>}
                                                      0

                                                          *
<switch term> ::= <switch factor> {AND <switch factor>}
                                                          0

                        1  ⎧ <switch id>      ⎫
<switch factor> ::= {^}    ⎨ (<switch expr>)  ⎬
                        0  ⎩ <relation>       ⎭

<relation> ::= <expression> { > } <expression>
                            { < }

<switch assignment> ::= <switch id> := <relation>

<switch id> ::= <id>

                                                        1
<macrocall> ::= <macro id> { ( <actual parm list> ) }
                                                        0

<macro id> ::= <id>

                                                         *
<actual parm list> ::= <actual parm> {, <actual parm>}
                                                         0

                   ⎧ <id>     ⎫
<actual parm> ::=  ⎨ <number> ⎬
                   ⎩ <int>    ⎭

                        ⎧ WRITE ⎫
<output statement> ::=  ⎨ PRINT ⎬ ( <output list> )
                        ⎩ PLOT  ⎭

                          1  ⎧                    4 ⎫
<output list> ::= { <int> ,} ⎨ <expr> {, <expr>}    ⎬
                          0  ⎪                    0 ⎪
                             ⎩ <id> ( <id> )        ⎭
```

Appendix B.

The content of the file SIL.HLP . All commands in the interactive editor and the graphic command system have in this file a 5 to 10 line description.

```
INTRO
   Welcome to the SIL system. Hope You will find it easy to use.
   You can run it in three different modes: 1) Batch mode,
   2) Graphic mode, and 3) Interactive mode dependent on the way
   you call the system (see the file SILREAD.ME).  In either
   graphic mode or interactive mode you may type several commands
   on the same line; just use a ; (semicolon) as separator between
   them. Together with the SIL system, you will find a small col-
   lection of sample problems - they should run without any diffi-
   culty. Also, they contain examples on how to use most of the
   options available in the system (MAXCPU, METHOD, STEPSIZE, DEBUG
   etc.). The  HELP  command is assumed to give sufficient assist-
   ance to cope with most problems. Stopping a 'wild' run can nor-
   mally be done by typing  S  or  s .

TOP
   No parameters.
   This command will cause the line pointer (actual line) to be
   moved to a virtual line before the first line of the file.

BOTTOM
   No parameters.
   The line pointer is moved to a virtual line following the last
   line of the file. This virtual line then becomes the actual
   line. In COMPILE mode, the rest of the model will be analyzed;
   if any errors are found, the editor will stop moving and display
   the line containing the error (or perhaps the following line).

INPUT  <parameter>
   Optional parameter: <nr>
   This command puts the editor in INPUT mode; that is, subse-
   quent lines are inserted in the file just after the actual
   line.  This mode is exited for one of three reasons.....
      1: an empty line (not a blank line) is entered
      2: if in COMPILE mode an erroneous statement is typed
      3: if INPUT<nr> ,  maximum <nr> lines can be inputted.
   Note: In the command  INPUT<nr> ,  <nr> can be any positive
      number and blanks may be inserted. Legal commands are:
      I; INP5; INPUT 22;    (*  commands are separated by ; *)
   An empty command (ENTER) is equvalent to INPUT.
```

ABORT

No parameters.
This command immediately terminates the edit session without updating the file on the disk. This is mainly for use if everything seems to go wrong. See also END.

END

No parameters.
Terminates the edit session. The model is analyzed by SIL, and if it is correct, then it is stored on disk and the model is run by the processing system. If the model is not correct, SIL will display the line with the error (or the following line) as the actual line and terminate the END command without storing anything; thus, it is impossible to leave the editor via the END command with an incorrect model. See the command ABORT. When issued from the graphic command system END terminates the command session. See the EDIT command.

HELP

Gives assistance. HELP HELP prints this message.

SET

Used for selecting between different modes of the compiler.
Valid commands are:

1: SET COMPILE ON	Causes the editor to make a SIL analysis of everything above and including the actual line.
2: SET COMPILE OFF	Do not analyse the model before the END command. Default.
3: SET LIST ON	Forces the editor to list on the terminal the lines treated by commands like COMPILE 10,15 or JUMP+5 .
4: SET LIST OFF	Types on the terminal only the actual line immediately after a command or a sequence of commands on one line. Default.

JUMP <parameter>

Optional parameters: <nr>| +<nr>| -<nr>|
This command without a parameter is equivalent to +1 or just 1. The editor will make the next line of the model the actual line. If a parameter is given, then the line with a relative line number (relative to the actual line) equal to this value is picked as the actual line. For convenience, JUMP+5 is the same as +5 or just 5.

NEW <parameter>

Optional parameters: <nr>| +<nr>| -<nr>| <nr>,<nr>| ,<nr>| <nr>,
The lines in the specified range will be printed on the terminal one by one and the user is prompted for new replacement lines. An empty line will terminate the command immediately.

DELETE <parameter>

Optional parameters: <nr>| +<nr>| -<nr>| <nr>,<nr>| ,<nr>| <nr>,
The lines specified are deleted and the line preceeding the deleted lines becomes the current line. The lines are effectively deleted from the file and there is no recovery facility, therefore use this command with care.

LIST <parameter>
Optional parameters: <nr>| +<nr>| <nr>,<nr>| ,<nr>| <nr>,
The lines in the specified range are listed on the terminal and the last one becomes the actual line.

COMPILE <parameter>
Optional parameters: <nr>| +<nr>| -<nr>|
The lines specified are compiled by SIL. If NOT in COMPILE mode all the preceeding lines are also compiled without any notice to the user other than error messages if necessary. The last line becomes the actual line. If the compiler is in LIST mode the lines compiled will be printed on the terminal.

FIND <string>
Optional parameter: <string>
From the line following the actual line, the editor will search the file for the first location of the prescribed text. If no string is specified, then the string used in the last FIND or REPLACE command will be used. See REPLACE.

REPLACE
Parameters: <string1> <string2>
In the actual line, the first location of <string1> will be exchanged with <string2>. If no paramters are specified, then the parameters from the latest FIND or REPLACE command with parameters will be used. The sequence of commands

```
REPLACE/ABCD/EF/;FIND;REPL;F;R;F;R;F;R
```

will replace the text 'ABCD' with 'EF' up to 5 times but only one occurence per line because the FIND command advances one line. All characters except blank and semicolon can be used as string-delimiters. The following commands are legal and identical:

```
REPLACE LABCDLEFL;  RE4ABCD4EF4;R   /ABCD/EF
```

SIM (<parameter>)
Optional parameter: <nr> : <nr>
Performs the simulation. When no argument list is given, the default values for the integration interval (either (0:1) or as specified in the SIL program) will be used. In the case where the left point of the specified integration interval coincides with the right point of the previous specified integration interval, the simulation will use the solution at this point as initial values for the simulation. This command WILL NOT clear the screen; therefore, one may change initial values or parameter values and present all the results on the same graph.

```
Ex:   SIM                         (* performs simulation *)
      SIM(0:5.5);  SIM(5.5:17)    (* simulates from 0 to 17 *)
```

PLOT (<argument list>)
Parameter; <nr>, <ident> , <ident> , ...
<nr>, <ident> (<ident>)
This function defines new variables to be plotted during the following SIM commands. The <int>, argument is optional; if omitted a default value of 250 is assumed. The <arglist> argument defines the variables to be sampled and plotted in the next simulation. The variables (up to 5) must be seperated with a comma. Notice, that expressions in the argument list are NOT allowed. When the argument is of the form <id1>(<id2>) it indicates that <id1> should be plotted as

function of <id2>. Notice, that only one PLOT command can be active at any one time since it will overwrite the information from the previous PLOT command (if any).

Ex: PLOT(A); PLOT(850,A(X)); PLOT(347,X,Y,Z)

DRAW (<argument list>)
Parameter; <ident> , <ident> , ...
<ident> (<ident>)
With this function one can view the results of the preceeding simulation. All variables sampled can be plotted on the screen. The <argument list> takes the same form as in PLOT, except that <nr> cannot be given. SIL will always use the sample of a variable with most points. The DRAW function will clear the screen before the drawing is done, and furthermore, the screen will be erased by the next SIM command. Note: the DRAW command does not redefine the variables to be plotted by the next SIM command.

SHOW <parameter>
Optional parameter: <ident>
This command will display the current values of all variables defined in the model. For dynamic variables the initial value is also displayed. Using <parameter> as command will display the value of that variable (SHOW Y is normally written as Y). This form of the command can be extended to <ident> , <ident> , <ident> ..... .
It is possible to change the value of a PARAMETER , or the initial value of a dynamic VARIABLE by a statement of the form <ident> := <value> .

XAXIS (<argument list>)
Parameter: <value> : <value> , <string>
<value> : <value>
, <string>
This function changes characteristics of the xaxis. The argument <value> : <value> will set the range of the axis as specified. The axis is redrawn with new divisions; the drawing area will not be cleared until the next DRAW or SIM command. The <string> parameter is used for labelling the axis; note, the first) will be the end of the string!

YAXIS (<argument list>)
Parameter: <value> : <value> , <string>
<value> : <value>
, <string>
This function changes characteristics of the yaxis. The argument <value> : <value> will set the range of the axis as specified. The axis is redrawn with new divisions; the drawing area will not be cleared until the next DRAW or SIM command. The <string> parameter is used for labelling the axis; note, the first) will be the end of the string!

TITLE (<string>)
This function is used for entering a new title for the plot. Initially the title will be 'SIL plot of results' unless the SIL-model itself contains a title, which will be used then. The <string> parameter may contain any character but ')' or carriage return.

DUMP_VARS

This command will immediately append a printout of the values of all sampled variables to the <fn>.LST file. The format will be as if the model had WRITE statements for all the variables.

FULLSCREEN

This command toggles the use of the screen between one full graphic screen with a seperate input/output screen and one screen having both the graphic and the input/output fields.

CLEAR

After this command the following SIM command will always clear the graphic screen before the new solution is displayed. This clear option is automatically invoked by the DRAW, PLOT, XAXIS, and YAXIS commands and is always active in 'fullscreen' mode.

EDIT

No parameters.
This command will terminate the graphic command system and return to editing the model. In case SIL has been invoked with the (Interactive parameter the system will go into edit mode initially.
See also the END command.

STOP

Not a legal command. Use END, EDIT, or ABORT instead.
In case you want to stop a 'wild' run, press the S or s key.

QUIT

Not a legal command. Use END, EDIT, or ABORT instead.

DEVICE

This is not a command, but a variable. In graphic mode SIL will attempt to autodetect the graphic board. In case your PC has multiple boards, SIL may not get the one you prefer. Generally, SIL will prefer high resolution to several colours. You may change the device used by changing the variable DEVICE as any variable is changed in graphic mode (DEVICE := 7). The following values are valid:
1: CGA, 2: MCGA, 3: EGA, 4: EGA64, 5: EGAMONO,
7: Hercules, 8: ATT400(Olivetti), 9: VGA, 10: PC3270.

Appendix C.

The list of error messages issued during compilation of the model or during the solution phase. The messages are numbered (not continuously) from 1 to 70. The errors from 54 to 70 are primarily issued by the runtime system. Some of the errors from 1 to 54 may be issued by the graphic command system. In the graphic command system only the error number is normally displayed, but the error message can be printed on the screen by the command HELP <number> . In case an error starting with COMPILER ERROR is encountered, the user is kindly requested to send all necessary documentation (primarily the model) on a diskette to the distributer.

1: COMPILER ERROR .. STACK OVERFLOW
An internal stack for temporary pointers has overflowed.

2: COMPILER ERROR .. I.P. AREA OVERFLOW
Is not likely to happen.

3: COMPILER ERROR .. TABLE OVERFLOW
An internal table is filled up; not likely to happen.

4: COMPILER ERROR .. SEVERE INTERNAL ERROR
The system has detected an internal lack of consistency in the generated internal code.

5: FURTHER MESSAGES SUPPRESSED
The μSIL system will not print more than 35 error messages in the .LST file. This message will be the last one, indicating that additional messages are suppressed.

6: INTERNAL RUN-TIME ERROR .. CONTACT SIL SUPPORT
This error indicates that the integrator reverses its direction of integration. It may indicate that one of the switch conditions behaves badly; that is, the solution is forced to follow the trajectory where the switch changes.

10: SCANNER ERROR .. SEVERE INTERNAL ERROR
Something is wrong internally; it is detected by the lexical scanner when converting the input model to sequences of symbols. One possible reason for this error is that the file SIL.ACC is changed unauthorized.

11: INCORRECT CONSTANT (OUT OF RANGE)
A constant in the model is not a legal number in the range of numbers on the actual computer. The demo version of the μSIL system has a smaller number range than other versions.

12: ILLEGAL CHARACTER: " " (DELETED)
The source of the model contains a character not recognized by the μSIL system. The character is effectively deleted.

13: MISSING FINAL "." (PERIOD)
The last symbol in a model must be a period "." . In case there is an error in the BEGIN END structure, the model may seem to end before it really does. Check the BEGIN END counter in the listing of the model.

14: INCORRECT OPERAND
An operand in <expression> is of wrong type. A variable of SWITCH type (say) cannot (correctly) appear in a calculation, this situation will cause this message.

15: TOO MANY NAMES
The μSIL compiler has an (installation dependent) bound on the number of variables that can be declared. Inserting the value of parameters into the model makes it possible to declare additional variables of other types.

16: TOO MANY CONSTANTS
The table used for storing the values of all the constants in the model is full. Reduce the number of constants either by declaring them as parameters or by using temporary variables for frequently occuring sub-expressions.

17: TOO MUCH MEMORY REQUIRED FOR MODEL
The model, in internal representation, requires more memory than is available on the computer. Reduce the model or increase the memory if possible.

20: INCORRECTLY FORMED DECLARATION OF VARIABLE abcd
The declaration of the variable 'abcd' is not syntactically correct. See section 6.2.1.3 VARIABLE for the correct syntax.

21: INCORRECTLY FORMED DECLARATION OF DERIVATIVE abcd
The declaration of the derivative 'abcd' is not syntactically correct. See section 6.2.1.4 DERIVATIVE for the correct syntax.

22: INCORRECTLY FORMED DECLARATION OF PARAMETER abcd
The declaration of the PARAMETER 'abcd' is not syntactically correct. See section 6.2.1.1 PARAMETER for the correct syntax.

23: INCORRECTLY FORMED DECLARATION OF SWITCH abcd
The declaration of the SWITCH 'abcd' is not syntactically correct. See section 6.2.1.5 SWITCH for the correct syntax.

24: INCORRECTLY FORMED DECLARATION
A declaration is syntactically wrong. There are three reasons for the error. Either an <identifier> was not found where it should be, or a range specification is incorrect, or either a TIME or SAMPLETIME declaration is incorrect.

25: MORE THAN ONE DECLARATION OF abcd
The variable 'abcd' is declared more than once.

26: MISSING INITIAL VALUE FOR abcd
The variable 'abcd' should have had an initial value associated before it is used for example in a DERIVATIVE declaration.

27: INCORRECT DECLARATION OF FORMAL PARAMETER
In a MACRO declaration the specification of the formal parameters is incorrect.

30: "abcd " IS UNDEFINED
The identifier 'abcd' is not previously defined. The reason could be either an error in the declaration or the name of the variable is mistyped. The error will also appear in case a variable is used as a global variable in a MACRO.

31: INCORRECT OPERAND TYPE FOR **
It is required that the exponent is an integer number, neither a real number nor a variable can be used. For real exponents use the functions EXP and LOG.

32: ASSIGNMENT INCOMPATIBLE WITH PREVIOUS ASSIGNMENT
In an assignment, the variable being assigned has a status for which the attempted assignment is invalid.

33: MISSING ASSIGNMENT OF abcd
The variable 'abcd' appears in an expression before it is assigned a value (one way or the other). Assigning initial values to dynamic variables is an assignment; parameters are always assigned as is the model time.

34: abcd PREVIOUSLY ASSIGNED A VALUE
The variable 'abcd' appears more than once on the left hand side of := . Normally, a variable can only be assigned once; the exception being assingments in connection with IF THEN ELSE constructions.

35: abcd ALSO HAS TO BE ASSIGNED IN "THEN"-BLOCK
If the variable 'abcd' is assigned in the ELSE part of an IF THEN ELSE statement, it must also be assigned in the THEN part of the same statement.

36: abcd ALSO HAS TO BE ASSIGNED IN "ELSE"-BLOCK
The variable 'abcd' is assigned in the THEN part of an IF THEN ELSE statement; it must also be assigned in the ELSE part of the statement.

37: MISMATCHED PARAMETER
When calling a MACRO there is not agreement between the type of an actual parameter and a formal parameter. This error also appear if either the TIME variable or a PARAMETER variable is in an output statement.

38: INDEX MUST BE SIMPLE VARIABLE OR <INT> NUMBER
Is not in use yet.

39: UPGRADING OF VARIABLE abcd NOT ALLOWED
Upgrading of an explicit static variable (say) occurs in an IF THEN ELSE statement when it is assigned a constant value (say) in the THEN part and a value depending on (say) an implicit static variable in the ELSE part. This is not allowed, the opposite is allowed though.

40: INCORRECT DIMENSION
Is not in use yet.

41: "<se2> " CANNOT FOLLOW "<se1> " HERE
In the source of the model the syntaxelement (token) <se2> follows <se1>. According to the syntax this is not allowed. Typically, this error will appear in case a ';' is missing.

42: BLOCK STRUCTURE MISMATCH
The end of the model file is reached before the END matching the initial BEGIN has been encountered; the number of BEGINs does not match the number of ENDs. Check the BEGIN END counter in the model listing.

43: EXPECTING RIGHT PARENTHESIS
In either an expression or a switch (logical) expression the number of left and right parentheses does not match. This error can also appear in connection with errorneous declarations.

44: EXPECTING < OR > BUT <se> WAS FOUND
In a <relation> the first <expression> is not terminated by either < or > ; this could be due to an error in the structure of the parentheses. The error only appears in either a switch assignment or in the switch expression between IF and THEN .

45: WRONG PROGRAM STRUCTURE
Normally appears together with either error number 42 or error number 13.

46: IMPLICIT ALGEBRAIC EQUATION MISMATCH IN IF-THEN-ELSE
The number of implicit algebraic equations in the THEN part of an IF THEN ELSE construction does not match the number of implicit algebraic equations in the ELSE part. It is a requirement in μSIL that the number of equations does not change in the different states of the model.

47: IMPLICIT DERIVATIVE EQUATION MISMATCH IN IF-THEN-ELSE
The number of equations for the implicitly given derivatives in a THEN part differs from the number of implicitly given derivative equations in the ELSE part. Similar to error number 46.

48: VARIABLE MISMATCH IN IMPLICIT EQUATION
This error will be issued in case an explicit derivative (say) appears in an implicit static equation. In general, this error indicates that an internal sequencing of the equations is impossible.

50: NR OF IMPLICIT VARIABLES DOES NOT MATCH NR OF EQUATIONS
SIL requires that the number of implicit static variables in a model matches the number of implicit static equations; this error says that this rule is violated.

51: NR OF IMPLICIT DERIVATIVES DOES NOT MATCH NR OF EQUATIONS
The number of implicit derivatives must be equal to the number of implicit derivative equations.

52: UNABLE TO FULFILL LANGUAGE REQUEST - INTERPRETER ASSUMED
The standard version of μSIL keeps a copy of the model in internal representation; during the solution of the model, this copy is interpreted. Other versions of the system can generate either a PASCAL, ALGOLW or FORTRAN program which is the copy of the model. If this option is not available, an attempt to use it will result in this error message.

53: MISSING OUTPUT SPECIFICATION

A model is not complete if there is no specification of how the output should be produced. At least one WRITE, PRINT or PLOT statement must be in the model.

54: TOO MANY OUTPUT POINTS

The system has a built-in limit on the total number of output points. Normally, this limit is so high that the size of the memory on the system determines the number of output points. The demo version has a severe limit on the number of output points.

55: TOO MANY OUTPUT TABLES

There are more than 10 output statements in the model. Since each output statement can have 5 variables, at most 50 variables in a model can be 'outputted'.

56: NO SAMPLES AVAILABLE FOR abcd

In the graphic command system the variable 'abcd' appears as argument for the DRAW command even though the variable does not appear in any output statement. The error is also issued in case no simulation has been performed before a DRAW command.

57: INVALID VALUE OF CONSTANT

In the graphic command system an invalid assignment of a system variable (METHOD, DEVICE etc.) is attempted.

60: RANGE CHECK ON abcd FAILED

The variable 'abcd' is declared in the model with a range. During the integration the value of the variable exceeds that range. Simulation is terminated.

61: INVALID SWITCH CONDITION ON abcd

The switch variable 'abcd' causes the simulation to a stop, probably due to an inconsistency in its definition. The solution is forced to follow a trajectory for which the switch is more or less undetermined.

62: TOO MANY <type> VARIABLES
The model defines too many variables of type <type>. The standard version of μSIL has rather high limits on all types of variables, whereas the demo version is rather limited with respect to this.

63: TOO MANY ITERATIONS REQUIRED WHEN SOLVING IMPLICIT EQNS
During the solution of the implicit static equations too many iterations have been used. The default limit is 250 but it can be changed by assigning the maximum to the variable MAXITER. This error implies either that there is no solution to the implicit algebraic equations (model error) or that the error requirement regarding the round off error level of the arithmetic is too strict. A better start guess for the iteration process, a limit on the stepsize (MAXSTEPSIZE) or a relaxation of the error bound can be tried in case there is no errors in the model equations.

64: TOO MUCH CPU-TIME REQUIRED
The simulation process is terminated with this error when more than MAXCPU seconds have been used. MAXCPU = 0 will turn off this check.

65: ITERATION MATRIX IS NUMERICALLY SINGULAR
When solving the system of implicitly given static equations using the Quasi-Newton algorithm, the linear equation solver discovered the singularity of the matrix. A possible reason for this singularity is a linear dependency between some of the equations.

66: SIMULATION TERMINATED BY USER INTERRUPT
A simulation is terminated when either of the keys S or s is pressed. This works both in the graphic command system and when running the system in batch mode.

67: LOGARITHM ARGUMENT <= 0
The LOG function is called with a negative argument. This can happen even if the argument is 'guaranteed' to be positive via a switch expression because the model normally is evaluated 'behind' a discontinuity.

68: DIVISION BY 0 (ZERO)

The runtime system catches attempts to divide by (exact) 0.

69: STEPSIZE TOO SMALL; PROBABLY CAUSED BY MODEL ERROR

An error in the model can force the integrator stepsize to be so small that it cannot significantly advance the model time. This stops the simulation.

Besides these error messages the user may see some initial error messages if the system is prevented from proper function. In case any of the files SIL.EXE, SIL.HLP, SIL.ERM and SIL.ACC is not present on the current disk drive or in the directory specified in the SIL.DEV file a message will be printed on the terminal and the system stops. This will also be the case if the model file (<name>.SIL) is missing in a non-editing mode of the system.

Since it is impossible to verify a large system as the μSIL system 100% one may accidentially see PASCAL error messages. If such errors happen in graphic mode, text mode can be entered by the DOS command 'MODE MONO'. Run the problem again but append a '> DUMP' to the command invoking the μSIL system. Copy the DUMP file, the model file, the .LST file and a description of the graphic command sequence used onto a diskette and mail it to the distributor.

Appendix D.

The documentation of the integration routine STRIDE. Taken from the original ALGOLW version.

D2 ORDINARY DIFFERENTIAL EQUATIONS
STRIDE ALGOLW VERSION 1 1979.08.24
LØSER BEGYNDELSES VÆRDI PROBLEMET FOR ET SÆT AF 1.ORDENS SÆDVANLIGE DIFFERENTIALLIGNINGER. KAN LØSE SÅVEL STIVE SOM IKKE-STIVE SYSTEMER. ANVENDER VARIABEL ORDEN - VARIABEL SKRIDTLÆNGDE SINGLY-IMPLICIT RUNGE-KUTTA MED FEJLESTIMAT.

1. FUNCTION.

THE PROCEDURE STRIDE SOLVES AN INITIAL VALUE PROBLEM FOR A SYSTEM OF SIMULTANEOUS 1.ORDER ORDINARY DIFFERENTIAL EQUATIONS.

Y' = F(X,Y)   X IN R AND Y,Y' IN R**N   (1)

2. USE.

THE PROCEDURE IS CALLED IN THE FOLLOWING WAY

```
STRIDE(N,F,JAC,VALUES)
```

WHERE

N   IS AN INTEGER HOLDING THE NUMBER OF SIMULTANEOUS EQUATIONS

F   IS A PROCEDURE COMPUTING THE RIGHT-HAND SIDE OF (1) AND DECLARED AS

```
PROCEDURE F(LONG REAL VALUE X;
            LONG REAL ARRAY Y,DY(*))
```

THE N VALUES EVALUATED IN THE POINT (X,Y) MUST BE STORED IN ARRAY DY.

JAC IS EITHER A DUMMY PROCEDURE (E.Q. F) OR A PROCEDURE THAT EVALUATE THE JACOBIAN MATRIX J OF THE SYSTEM. ELEMENT J(R,S) BEING DEFINED AS

```
            DFR(X,Y)
J(R,S) = ----------  .
              DYS
```

THE PROCEDURE IS DECLARED AS

```
PROCEDURE JAC(LONG REAL VALUE X;
              LONG REAL ARRAY Y(*);
              LONG REAL ARRAY JACOBIAN(*,*))
```

THE N*N VALUES EVALUATED IN THE POINT (X,Y) MUST BE STORED IN THE ARRAY JAKOBIAN.

VALUES IS A PROCEDURE USED FOR THE 'USER-STRIDE' COMMUNICATION. INITIAL VALUES MUST BE GIVEN IN VALUES AND ON REQUEST IT IS CALLED WHEN OUTPUT-VALUES ARE AVAILABLE. IT IS DECLARED AS

```
PROCEDURE VALUES(LONG REAL VALUE RESULT X, XPRINT;
                 LONG REAL ARRAY Y, IMAG(*);
                 INTEGER   VALUE RESULT INDEX;
                 LONG REAL VALUE RESULT STP, STPMAX;
                 INTEGER   VALUE RESULT M;
                 LONG REAL ARRAY EPS(*);
                 INTEGER   VALUE RESULT METHOD;
                 INTEGER   ARRAY LIMITS(*))
```

AND MUST HAVE THE STRUCTURE

```
IF  INDEX = 1  THEN
    INITIALIZE X, XPRINT, Y(1-N), EPS(1-N)
    AND EVENTUALLY
    YMAG(1-N), STP, STPMAX, M, METHOD AND
    LIMITS(1-N)
ELSE
    PRINT RESULTS (X, Y(1-N), INDEX, STP, M)
    AND EVENTUALLY
    ADJUST XPRINT, YMAG(1-N), STPMAX, EPS(1-N),
    METHOD AND LIMITS(1-N).
```

THE SIGNIFICANCE OF THE PARAMETERS IS THE FOLLOWING

INDEX DETERMINS FOR WHAT REASON VALUES HAS BEEN CALLED ACCORDING TO THE FOLLOWING TABLE

INDEX = 1. INITIALIZATION. X AND Y(1-N) MUST BE SET TO THEIR INITIAL VALUES. FURTHERMORE XPRINT MUST BE SET TO THE FIRST OUTPUT-POINT AND IN EPS(1-N) THE N BOUNDS ON THE LOCAL ACCURACY ON THE COMPONENTS OF Y(1-N) SHOULD BE STORED. THE VARIABLES YMAG(1-N), STP, STPMAX, M, METHOD AND LIMITS(1-N) MAY HAVE THEIR DEFAULT VALUES CHANGED.

= 2. OUTPUT AT A PREVIOUSLY SPECIFIED POINT. (X,Y) HAVE BEEN SET TO THE SOLUTION WITH X EQUAL TO A PREVIOUSLY SELECTED VALUE OF XPRINT. XPRINT HAS BEEN UPDATED TO ACHIEVE EQUALY SPACED OUTPUT-POINTS.

= 3. THE END OF A STEP HAS BEEN REACHED. (X,Y) IS THE SOLUTION AT THIS POINT. THE PARAMETER LIMITS (SEE BELOW) MAY BE USED TO AVOID THE CALLS OF VALUES WITH INDEX = 3 (OR INDEX>3).

= 4. STEP NOT COMPLETED.
DIVERGENCE OF ITERATION PROCEDURE IN METHOD. INTEGRATION RESUMES FROM LAST SUCCESFULL STEP.

```
= 5. STEP NOT COMPLETED.
     FAILURE TO CONVERGE WITIN ALLOWED NUMBER OF
     ITERATIONS (SEE METHOD). INTEGRATION RESUMES
     FROM LAST SUCCESFULL STEP.

= 6. STEP NOT COMPLETED.
     LOCAL ERROR ESTIMATE GREATER THAN ALLOWED
     VALUE. INTEGRATION RESUMES FROM LAST SUCCES
     FULL STEP.

= 7. WARNING.
     NEXT VALUE OF XPRINT IS ON WRONG SIDE OF THE
     PRESENT STEP, THE INTEGRATION DIRECTION WILL
     BE REVERSED.

  THE VALUE OF INDEX SHOULD NEVER BE ALTERED BY
  VALUES, EXCEPT IN ONE CIRCUMSTANCE. IF FOR SOME
  REASON THE USER WANTS THE INTEGRATION STOPPED
  BEFORE NORMAL TERMINATION, HE MAY PUT INDEX = 0.
  NORMAL TERMINATION IS OBTAINED BY MEANS OF
  LIMITS (SEE BELOW).

=====>    METHOD  DETERMINES WHICH STRIDE-OPTIONS TO BE USED
               FOR THE SOLUTION OF A GIVEN PROBLEM. METHOD IS A
               3-DIGIT NUMBER COMPOSED BY

                METHOD = 100 * MERPST + 10 * MTYPE + MAXITS

               WHERE MERPST=0 OR 1, MTYPE=0,1,2 OR 3 AND
               MAXITS=1,2,...,9 WITH THE FOLLOWING SIGNIFICANCE

               MERPST = 0. LOCAL ERROR PER UNIT STEP ABSOLUTE
                     ACCURACY CONTROL IS TO BE USED.

                     = 1. LOCAL ERROR PER STEP ABSOLUTE ACCURACY
                       CONTROL IS TO BE USED.

               MTYPE  = 0. FIXED POINT ITERATION WITH FIXED
                    NUMBER OF ITERATIONS IS TO BE USED.

                     = 1. FIXED POINT ITERATION WITH A LIMIT ON
                       THE NUMBER OF ITERATION IS TO BE USED.

                     = 2. NEWTON ITERATION WITH A USER SUPPLIED
                       JACOBIAN.

                     = 3. NEWTON ITERATION WITH A JACOBIAN
                       EVALUATED BY DIVIDED DIFFERENCES.

               MAXITS = P. THE NUMBER OF ITERATIONS IN EACH
                    STEP IS (LIMITED BY) P.

               NO OTHER VALUES ARE ALLOWED. RECOMMENDED VALUES
               ARE  MERPST = 1. MTYPE = 0 OR 1 FOR NON-STIFF
               PROBLEMS,  MTYPE = 2 OR 3 FOR STIFF PROBLEMS AND
               MAXITS <= 3.
               DEFAULT VALUE IS METHOD = 101 (NON-STIFF
               SYSTEMS).
```

```
LIMITS  AN INTEGER ARRAY OF LENGHT 7. THE NUMBERS RE-
     PRESENT LIMITS ON STRIDE-ACTIONS AS SHOWN IN THE
     TABLE BELOW.  THE NUMBER P MAY BE CHANGED ON
     EVERY CALL OF VALUES.

!   I       LIMITS(I) = 0         !   LIMITS(I) = P > 0
!   1  NO LIMIT ON TOTAL NUM-    ! TOTAL NUMBER OF STEPS
!      BER OF STEPS.             ! LIMITED TO P.
!   2  NO LIMIT ON TOTAL NUM-    ! TOTAL NUMBER OF OUTPUT
!      BER OF OUTPUT POINTS.     ! POINTS LIMITED TO P.
!   3  NO CALL OF VALUES WITH    ! VALUES ARE CALLED ONLY
!      INDEX = 3                 ! AFTER P COMPLETED
!                                ! STEPS.
! 4-7 NO CALL OF VALUES WITH     ! VALUES ARE CALLED WITH
!      INDEX = I (4,5,6 OR 7).   ! INDEX = I ONLY AFTER
                                 ! P FAILED STEPS WITH
                                 ! THIS KIND OF FAILURE.

     THE DEFAULT VALUES ARE (0,1,0,0,0,0,0). ON AN
     INDEX = 2 CALL ONE MIGHT GET SOME INFORMATION
     ABOUT HOW STRIDE ACTED ON THE ACTUAL SOLUTION
     (SEE TABLE BELOW).

     I          P = -LIMITS(I) - 1.

     1     P IS THE TOTAL NUMBER OF INITIATED STEPS.

     2     P IS THE TOTAL NUMBER OF OUTPUT POINTS.

     3     P IS THE TOTAL NUMBER OF ACCEPTED STEPS
             SINCE THE LATEST INDEX = 3 CALL OF
             VALUES.

     4-7   P IS THE TOTAL NUMBER OF STEPS FAILED SINCE
             THE LATEST INDEX = I CALL OF VALUES.

X          THE INDEPENDENT VARIABLE (OFTEN THE TIME)
     WITH SIGNIFICANCE ACCORDING TO THE VALUE OF
     INDEX.

     INDEX
       = 1 : X MUST BE SET AT AN INITIAL VALUE.

         = 2 : X CONTAIN THE VALUE OF AN OUTPUT POINT.

         = 3 : X IS THE END OF A SUCCESFULLY COMPLETED
                 STEP.

         > 3 : X IS THE BEGINNING OF A FAILED STEP.

Y          CURRENT VALUE OF THE DEPENDENT VARIABLE VEC-
     TOR (SOLUTION TO THE DIFFERENTIAL EQUATION) COR-
     RESPONDING TO THE VALUE OF X. MUST BE INITIA-
     LIZED ON AN INDEX = 1 CALL.
XPRINT  THE NEXT VALUE OF X FOR AN INDEX = 2 CALL
     (OUTPUT). AT INDEX = 1 IT SHOULD BE GIVEN A
     VALUE AND AS DEFAULT OUTPUT POINTS ARE EQUALLY
     SPACED. THIS MAY BE CHANGED AT ANY VALUES CALL
     (NORMALLY ON INDEX = 2 CALLS).
```

YMAG A VECTOR CONTAINING IN ITS I'TH COMPONENT AN INDICATION OF THE VALUE OF !Y(I)! OVER RECENT STEPS. STRIDE DOES NOT USE THE INFORMATION IN YMAG, BUT IT MIGHT BE USED BY THE USER IF HE/SHE WANTS RELATIVE ERROR CONTROL RATHER THAN ABSOLUTE (SEE EPS).

=====> STP THE STEPSIZE. ON AN INDEX = 1 CALL IT MAY BE SET BY THE USER, BUT IF NOT SET STRIDE WILL ESTIMATE ONE ITSELF. ON ALL OTHER CALLS STP GIVES THE CURRENT VALUE OF THE STEPSIZE.

=====> STPMAX A USER SUPPLIED BOUND ON THE MAGNITUDE OF STP. IT MAY BE SET OR CHANGED ON ALL CALLS TO VALUES. IF STPMAX = 0 (THE DEFAULT VALUE) NO BOUND IS IMPOSED.

=====> M THE ORDER. ON AN INDEX = 1 CALL THE MAXIMUM ORDER OF THE METHOD (1-15) MAY BE SET IN M IF THE DEFAULT VALUE 10 IS NOT APPROPRIATE. ON ALL OTHER CALLS M IS THE CURRENT VALUE OF THE ORDER.

EPS A VECTOR CONTAINING IN ITS I'TH COMPONENT A BOUND IN ABSOLUT VALUE ON THE LOCAL ACCURACY OF THE I'TH COMPONENT OF Y. MUST BE INITIALIZED ON AN INDEX = 1 CALL AND MAY BE CHANGED ON ALL OTHER CALLS IF THERE IS A REASON FOR IT. RELATIVE ERROR CONTROL MAY BE ESTABLISHED BY

EPS(I) = EPSREL(I) * YMAG(I), I=1,N

WHERE EPSREL(I) IS THE RELATIVE BOUND ON THE I'TH COMPONENT. THIS WORKS BEST, IF LIMITS(3) = 1 THUS CHANGING EPS AFTER EACH STEP.

3. TIME AND STORAGE REQUIREMENTS.

THE TIME REQUIRED DEPENDS ON THE ACTUAL PROBLEM TO BE SOLVED AND THE ACCURACY SPECIFIED. THE STORAGE USED IS ABOUT 20 KBYTES ON THE NEUCC 3033 EXPRESS ALGOLW VERSION NOT COUNTING THE ARRAYS.

4. METHOD.

STRIDE IS AN IMPLEMENTATION OF THE SINGLY-IMPLICIT RUNGE-KUTTA METHODS DESCRIBED IN BUTCHER(1). THEY ARE IMPLEMENTED AS A VARIABLE ORDER VARIABLE STEPSIZE SCHEME WITH 15 AS MAXIMUM ORDER. THE PROCEDURE IS WELL-SUITED BOTH FOR STIFF SYSTEMS AND FOR NON-STIFF SYSTEMS. THE RELATIVE ACCURACY REQUIREMENTS MAY VARY FROM ALMOST NOTHING TO ORDER THOUSAND UNITS OF MACHINE ACCURACY BUT COSTS GROWS WITH HIGHER ACCURACY.

5. DISCUSSION.

THE PROGRAM SHOULD NOT BE USED FOR SOLVING LARGE (N > 100) STIFF SYSTEMS WITH A SPARSE JACOBIAN. SOLVING THE SYSTEMS OF LINEAR EQUATIONS INVOLVED IS PERFORMED BY MEANS OF ORDINARY GAUSS ELIMINATION WITHOUT SPARSE MATRIX TECHNIQUE.

6. EXAMPLE.

SOLVE Y' = -Y , Y(0) = 1 , X IN THE RANGE (0,16) , LOCAL ACCURACY 10'-4 IN ABSOLUTE VALUE AND WITH OUTPUT AT X = 1, 2, 4, 8 AND 16. THE PROBLEM IS SOLVED BY THE FOLLOWING PROGRAM

```
BEGIN
%STRIDE
PROCEDURE F(LONG REAL VALUE X;
            LONG REAL ARRAY Y, DY(*));
DY(1):=-Y(1);
PROCEDURE VALUES(LONG REAL VALUE RESULT X, XPRINT;
                 LONG REAL ARRAY Y, YMAG(*);
                 INTEGER  VALUE RESULT INDEX;
                 LONG REAL VALUE RESULT STP, STPMAX;
                 INTEGER  VALUE RESULT M;
                 LONG REAL ARRAY EPS(*);
                 INTEGER  VALUE RESULT METHOD;
                 INTEGER  ARRAY LIMITS(*));
IF INDEX = 1 THEN
   BEGIN
   X:=0;
   XPRINT:=1;
   Y(1):=1;
   EPS(1):='-4L;
   LIMITS(2):=5;
   METHOD:=115;
   WRITE("     X             Y(1)          ERROR");
   R_W:=12; S_W:=1; R_FORMAT:="S"
   END
ELSE
   BEGIN
   WRITE(X,Y(1),LONGEXP(-X)-Y(1));
   XPRINT:=X+X
   END;
STRIDE(1,F,F,VALUES)
END.
```

THE PROGRAM PRODUCE THE FOLLOWING OUTPUT

```
           X            Y(1)         ERROR
  1.0000'+00    3.6778'-01    1.5250'-05
  2.0000'+00    1.3539'-01   -5.6492'-05
  4.0000'+00    1.8330'-02   -1.4794'-05
  8.0000'+00    2.8103'-04    5.4436'-05
  1.6000'+00   -2.6745'-05    2.6845'-05
```

7. REFERENCE.

(1) J. BUTCHER : ON THE IMPLEMENTATION OF IMPLICIT RUNGE-KUTTA METHODS. BIT 16 (1976), P. 237-240

FOR FURTHER INFORMATION CONTACT P. G. THOMSEN , NUMERISK INSTITUT, DTH. TEL. 02-881911 LOK. 4373.

LIST OF REFERENCES.

Burrage K., Butcher J.C., Chipman F.:
An implementation of Singly Diagonally Implicit Runge-Kutta Methods.
BIT, 1980, vol. 20, pp. 326-340.

Elmquist H.:
SIMNON. An Interactive Simulation Program for Non-linear systems. User's manual.
Dept. of Automatic Control, Lund Institute of Technology, Sweden. 1975, Report 7502.

Drønen O.:
The Differential Equations of the Asynchronous Machine. Physical interpretation (in Danish).
Danish Engineering Academy. 1985.

Jennings A.:
Matrix Computations for Engineers and Scientists.
John Wiley and Sons, 1985.

Lambert J.:
Computational Methods in Ordinary Differential Equations.
Wiley and Sons. 1973.

Lewin P.A., Bjørnø L.:
Acoustically induced shear stresses in the vicinity of microbubbles in tissue.
Journal of the Acoustical Society of America, 1982,
vol. 71(3), pp. 728-734.

Moler C.:
MATLAB Users Guide.
Department of Computer Science, The University of New Mexico.
Technical report (1980).

Nielsen O.B.:
The Interaction of Liquid and Gaseous Fuels in the Dual Fuel Engine. (PhD theses).
Laboratory for Energetics, Technical University of Denmark.
Report RE 86-7, 1986.

Pritsker A.A.B.:
The GASP IV Simulation Language.
John Wiley and Sons. 1974.

Wirth N.:
Algorithms + Datastructures = Programs
Prentice-Hall 1976.

Index

Vol. 379: A. Kreczmar, G. Mirkowska (Eds.), Mathematical Foundations of Computer Science 1989. Proceedings, 1989. VIII, 605 pages. 1989.

Vol. 380: J. Csirik, J. Demetrovics, F. Gécseg (Eds.), Fundamentals of Computation Theory. Proceedings, 1989. XI, 493 pages. 1989.

Vol. 381: J. Dassow, J. Kelemen (Eds.), Machines, Languages, and Complexity. Proceedings, 1988. VI, 244 pages. 1989.

Vol. 382: F. Dehne, J.-R. Sack, N. Santoro (Eds.), Algorithms and Data Structures. WADS '89. Proceedings, 1989. IX, 592 pages. 1989.

Vol. 383: K. Furukawa, H. Tanaka, T. Fujisaki (Eds.), Logic Programming '88. Proceedings, 1988. VII, 251 pages. 1989 (Subseries LNAI).

Vol. 384: G. A. van Zee, J. G. G. van de Vorst (Eds.), Parallel Computing 1988. Proceedings, 1988. V, 135 pages. 1989.

Vol. 385: E. Börger, H. Kleine Büning, M. M. Richter (Eds.), CSL '88. Proceedings, 1988. VI, 399 pages. 1989.

Vol. 386: J.E. Pin (Ed.), Formal Properties of Finite Automata and Applications. Proceedings, 1988. VIII, 260 pages. 1989.

Vol. 387: C. Ghezzi, J. A. McDermid (Eds.), ESEC '89. 2nd European Software Engineering Conference. Proceedings, 1989. VI, 496 pages. 1989.

Vol. 388: G. Cohen, J. Wolfmann (Eds.), Coding Theory and Applications. Proceedings, 1988. IX, 329 pages. 1989.

Vol. 389: D.H. Pitt, D.E. Rydeheard, P. Dybjer, A.M. Pitts, A. Poigné (Eds.), Category Theory and Computer Science. Proceedings, 1989. VI, 365 pages. 1989.

Vol. 390: J.P. Martins, E.M. Morgado (Eds.), EPIA 89. Proceedings, 1989. XII, 400 pages. 1989 (Subseries LNAI).

Vol. 391: J.-D. Boissonnat, J.-P. Laumond (Eds.), Geometry and Robotics. Proceedings, 1988. VI, 413 pages. 1989.

Vol. 392: J.-C. Bermond, M. Raynal (Eds.), Distributed Algorithms. Proceedings, 1989. VI, 315 pages. 1989.

Vol. 393: H. Ehrig, H. Herrlich, H.-J. Kreowski, G. Preuß (Eds.), Categorical Methods in Computer Science. VI, 350 pages. 1989.

Vol. 394: M. Wirsing, J.A. Bergstra (Eds.), Algebraic Methods: Theory, Tools and Applications. VI, 558 pages. 1989.

Vol. 395: M. Schmidt-Schauß, Computational Aspects of an Order-Sorted Logic with Term Declarations. VIII, 171 pages. 1989. (Subseries LNAI).

Vol. 396: T.A. Berson, T. Beth (Eds.), Local Area Network Security. Proceedings, 1989. IX, 152 pages. 1989.

Vol. 397: K.P. Jantke (Ed.), Analogical and Inductive Inference. Proceedings, 1989. IX, 338 pages. 1989. (Subseries LNAI).

Vol. 398: B. Banieqbal, H. Barringer, A. Pnueli (Eds.), Temporal Logic in Specification. Proceedings, 1987. VI, 448 pages. 1989.

Vol. 399: V. Cantoni, R. Creutzburg, S. Levialdi, G. Wolf (Eds.), Recent Issues in Pattern Analysis and Recognition. VII, 400 pages. 1989.

Vol. 400: R. Klein, Concrete and Abstract Voronoi Diagrams. IV, 167 pages. 1989.

Vol. 401: H. Djidjev (Ed.), Optimal Algorithms. Proceedings, 1989. VI, 308 pages. 1989.

Vol. 402: T.P. Bagchi, V.K. Chaudhri, Interactive Relational Database Design. XI, 186 pages. 1989.

Vol. 403: S. Goldwasser (Ed.), Advances in Cryptology – CRYPTO '88. Proceedings, 1988. XI, 591 pages. 1990.

Vol. 404: J. Beer, Concepts, Design, and Performance Analysis of a Parallel Prolog Machine. VI, 128 pages. 1989.

Vol. 405: C.E. Veni Madhavan (Ed.), Foundations of Software Technology and Theoretical Computer Science. Proceedings, 1989. VIII, 339 pages. 1989.

Vol. 407: J. Sifakis (Ed.), Automatic Verification Methods for Finite State Systems. Proceedings, 1989. VII, 382 pages. 1990.

Vol. 408: M. Leeser, G. Brown (Eds.),Hardware Specification, Verification and Synthesis: Mathematical Aspects. Proceedings, 1989. VI, 402 pages. 1990.

Vol. 409: A. Buchmann, O. Günther, T.R. Smith, Y.-F. Wang (Eds.), Design and Implementation of Large Spatial Databases. Proceedings, 1989. IX, 364 pages. 1990.

Vol. 410: F. Pichler, R. Moreno-Diaz (Eds.), Computer Aided Systems Theory – EUROCAST '89. Proceedings, 1989. VII, 427 pages. 1990.

Vol. 411: M. Nagl (Ed.), Graph-Theoretic Concepts in Computer Science. Proceedings, 1989. VII, 374 pages. 1990.

Vol. 412: L.B. Almeida, C.J. Wellekens (Eds.), Neural Networks. Proceedings, 1990. IX, 276 pages. 1990.

Vol. 413: R. Lenz, Group Theoretical Methods in Image Processing. VIII, 139 pages. 1990.

Vol. 414: A.Kreczmar, A. Salwicki, M. Warpechowski, LOGLAN '88 – Report on the Programming Language. X, 133 pages. 1990.

Vol. 415: C. Choffrut, T. Lengauer (Eds.), STACS 90. Proceedings, 1990. VI, 312 pages. 1990.

Vol. 416: F. Bancilhon, C. Thanos, D. Tsichritzis (Eds.), Advances in Database Technology – EDBT '90. Proceedings, 1990. IX, 452 pages. 1990.

Vol. 417: P. Martin-Löf, G. Mints (Eds.), COLOG-88. International Conference on Computer Logic. Proceedings, 1988. VI, 338 pages. 1990.

Vol. 419: K. Weichselberger, S. Pöhlmann, A Methodology for Uncertainty in Knowledge-Based Systems. VIII, 136 pages. 1990. (Subseries LNAI).

Vol. 420: Z. Michalewicz (Ed.), Statistical and Scientific Database Management, V SSDBM. Proceedings, 1990. V, 256 pages. 1990.

Vol. 421: T. Onodera, S. Kawai, A Formal Model of Visualization in Computer Graphics Systems. X, 100 pages. 1990.

Vol. 423: L. E. Deimel (Ed.), Software Engineering Education. Proceedings, 1990. VI, 164 pages. 1990.

Vol. 424: G. Rozenberg (Ed.), Advances in Petri Nets 1989. VI, 524 pages. 1990.

Vol. 426: N. Houbak, SIL – a Simulation Language. VII, 192 pages. 1990.